Social Policy in Ageing Societies

Social Policy in Ageing Societies

Britain and Germany Compared

Edited by

Alan Walker
University of Sheffield, UK

Gerhard Naegele
University of Dortmund, Germany

palgrave
macmillan

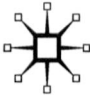

First published 2009 by
PALGRAVE MACMILLAN

Palgrave Macmillan in the UK is an imprint of Macmillan Publishers Limited,
registered in England, company number 785998, of Houndmills, Basingstoke,
Hampshire RG21 6XS.

Palgrave Macmillan in the US is a division of St Martin's Press LLC,
175 Fifth Avenue, New York, NY 10010.

Palgrave Macmillan is the global academic imprint of the above companies
and has companies and representatives throughout the world.

Palgrave® and Macmillan® are registered trademarks in the United States,
the United Kingdom, Europe and other countries.

ISBN-13: 978–0–230–52098–1 hardback
ISBN-10: 0–230–52098–7 hardback

This book is printed on paper suitable for recycling and made from fully
managed and sustained forest sources. Logging, pulping and manufacturing
processes are expected to conform to the environmental regulations of the
country of origin.

A catalogue record for this book is available from the British Library.

Library of Congress Cataloging-in-Publication Data
Social policy in ageing societies : Britain and Germany compared / [edited by]
    Alan Walker, Gerhard Naegele.
        p.   cm.
    ISBN 978–0–230–52098–1 (alk. paper)
    1. Older people—Great Britain.   2. Older people—Germany.   3. Older
    people—Cross-cultural studies.   4. Great Britain—Social policy.
    5. Germany—Social policy.   I. Walker, Alan.   II. Naegele, Gerhard,
    1948–
    HQ1064.G7S73 2008
    305.260941—dc22   2008030684

10   9   8   7   6   5   4   3   2   1
18 17 16 15 14 13 12 11 10 09

Printed and bound in Great Britain by
CPI Antony Rowe, Chippenham and Eastbourne

Contents

List of Figures and Tables

Figures

Tables

Contributors

Gertrud M. Backes, Dr.phil., Dipl.soziologin, is Professor and Director for Ageing and Society at Vechta University, Centre for Research on Ageing and Society and is also the speaker in the 'Ageing and Society' section of the German Sociological Association. She held a professorship of Sociology at the University of Applied Sciences in Cottbus from 1993 to 1998, a professorship of Gerontology: Sociology and Social Policy at Vechta University from 1998 to 2000, and more recently she worked on the topic of Social Gerontology at Kassel University from 2000 to 2006. Her research interests include gender and ageing, ageing, social change and social structure, and 'Lebenslagen': gender, body and ageing.

Ingrid Eyers is Lecturer in Health and Social Care in the Faculty of Health and Medical Sciences at the University of Surrey, and is a member of the Centre for Research on Ageing and Gender within the University. Her current research forms part of the NDA-funded research project SomnIA (Sleep in Ageing).

Uwe Fachinger is Professor of Economics and Demography at the Institute for Gerontology at Vechta University and Member of the Centre for Research on Ageing and Society, Vechta University. His research interests encompass economic analysis of social and distributional policy, on a national and international level.

Frerich Frerichs, Dr.phil.habil., is a sociologist, psychologist, and Professor of Ageing and Work in the Institute of Gerontology and is also a member of the Research Centre on Ageing and Society (CAS), both based at Vechta University. Until 2006 he was head of the Department of Demographic Change, Labour Market and Social Policy for older workers at the Institute of Gerontology in Dortmund. His current research activities encompass labour market / social policy for older workers and employment policies / human resource management for an ageing workforce.

Vera Gerling, Dr.phil., studied Sociology, Political Science and Psychology at the Rheinische Friedrich-Wilhelms Universität in Bonn and took her doctoral degree at the University of Dortmund in the Pedagogy and

Sociology departments. From 1996 to 2006 she was a research assistant, department head and most recently acting chief executive at the Institute of Gerontology at the University of Dortmund. From 2006 to 2007 she was project development leader at the Projektgesellschaft für Zukunfts- und Potenzialentwicklung in Lünen. Since the autumn of 2007 she has worked as a freelance social gerontologist for GER-ON Consult & Research in Dortmund.

Jay Ginn, PhD, was employed as a Senior Research Fellow in the Sociology Department of the University of Surrey and was a co-director of the Centre for Research on Ageing and Gender until her retirement in November 2004. She now works at the University of Surrey as a Visiting Professor. Her research interests include gender, class and ethnic differences in the economic resources of older people, and the impact on women of pension system reforms, cross-nationally, especially the shift from state to private pensions. Her recent books include *Women, Work and Pensions* (2001) (co-edited with D. Street and S. Arber) and *Gender, Pensions and Lifecourse*, published in 2003.

Caroline Glendinning is Professor of Social Policy and Assistant Director of the Social Policy Research Unit (SPRU) at the University of York. She leads SPRU's Department of Health funded research programme on Choice and Independence over the Lifecourse; and has contributed to the national evaluation of the Individual Budget Pilot Projects. She has longstanding research interests in social gerontology, health and social services, disability and informal care; and in the funding, organisation and delivery of long-term and social care services in other countries.

Tobias Gross, Diploma in Economics, MA in European Studies, is a Research Fellow at the European Policies Research Centre (EPRC), University of Strathclyde. His main work relates to regional policy and Structural Funds in Austria and Germany. Since 2008 he has been working for the Bank für Sozialwirtschaft.

Josef Hilbert, Dr.rer.soc., is a sociologist and Privatdozent at the Medical Faculty of the Ruhr, Bochum. He has been a Research Fellow at the Institute for Work and Technology – University of Applied Sciences, Gelsenkirchen since 1989, and is currently head of the Health Industries and Quality of Life research department. His fields of research include innovation in health and care industries, the economics of ageing societies, and vocational training for health and care businesses.

Gerhard Igl is Professor of Social Security Law at the University of Kiel, and Director of the Institute of Social Security Law and Social Policy in Europe. His main research interests include social security law in Europe and in the US, long-term care in Europe and the US, and the financing of medical and social services and institutions.

Tony Maltby, PhD, is CROW Research Fellow and Deputy Director at the Centre for Research into the Older Workforce at the National Institute for Adult Continuing Education (England and Wales), based in Leicester. His research interests embrace a range of subjects including work, income and the social policy of later life.

Gerhard Naegele is a Professor and holder of the chair in Social Gerontology and Director of the Institute of Gerontology at the Technical University of Dortmund. He is the author of numerous publications concerning social and social gerontological issues. His research interests are in social gerontology, social policy, older workers, social services and poverty.

Anita B. Pfaff, Dr.phil., is a Professor at the University of Augsburg. She works in the fields of micro-economic theory, distribution theory, and the theory and politics of job markets, with particular emphasis on socio-political issues and the socio-economics of health services. She has many years' scientific and political advisory experience, for example as a member of the Advisory Group on Women of the then Bundesministerium für Jugend, Familien, Frauen und Gesundheit (1989–93), of the Advisory Group on Area Planning of the Bundesministerium für Bauwesen und Städtebau (1991–94), and currently of the Commission of Enquiry into Demographic Change of the German Bundestag. She is deputy chairperson of the Scientific Advisory Group of the Deutsches Institut für Wirtschaftsforschung, co-founder and deputy director of the Internationales Institut für Empirische Sozialökonomie, and an executive committee member of the Bayerischer Forschungsverbund Public Health – Öffentliche Gesundheit.

Judith Phillips is Professor of Gerontology and Social Work at Swansea University and director of the Older People and Ageing Research and Development Network (OPAN Cymru) in Wales. Her research interests are in social work and social care and include carers in employment, housing and retirement communities, family and kinship networks, care work and older offenders. She is series editor for *Ageing and the Lifecourse* (Policy Press) and is President of the British Society of Gerontology.

Wolfgang Potratz, Dr. rer.soc., a political scientist, studied Political Sciences, Economic Policy, History and English Literature. He has been a Research Fellow at the Institute for Work and Technology, University of Applied Sciences, Gelsenkirchen since 1989 and is an external lecturer at Ruhr University in Bochum. His research fields include the comparative economics of health and care industries, work in health and care, and the economics of ageing societies.

Monika Reichert, Dr.phil., studied Psychology in Gieben and Cologne and took her doctoral degree at the Freie Universität in Berlin. From 1990 to 2000 she worked as a research assistant before becoming Scientific Chief Executive at the Institute of Gerontology at the University of Dortmund in 2000. She has worked as a Visiting Professor at institutes such as the Evangelische Fachhochschule in Berlin and the Social Gerontology and Geragogics department at the University of Dortmund. Since 2005 she has been Professor of Social Gerontology at the University of Dortmund. Her research interests include domestic nursing, social relationships in age, education in age, life circumstances and vitae.

Christiane Rohleder took her first and doctoral degrees at the Westfälische Wilhelmsuniversität in Münster. From 1992 to 1995 she worked as a research assistant in the Department of Pedagogy at the same institution. From 1996 to 2001 she worked as research assistant at the University of Dortmund and since 2001 she has been Professor of Sociology at the Katholische Fachhochschule Nordrhein-Westfalen in Münster. Her research interests include welfare activity with older people, and the sociology of families, especially in the area of domestic violence.

Philip Taylor is Professor of Employment Policy in the Faculty of Business and Enterprise, Swinburne University of Technology, Melbourne. His research interests include public policy towards older workers and the response of business to workforce ageing.

Winfried Schmähl, Dr.rer.pol., was full Professor of Economics (especially social policy) and Director of the Department of Economics, Centre for Social Policy Research, University of Bremen until July 2007.

His numerous publications have been mainly in the fields of social policy (especially financing) and old-age security. He is also a member of several advisory boards as has been Chairman of the Social Advisory Board on Pension Policy to the German federal government since 1994.

Alan Walker is Professor of Social Policy and Social Gerontology at the University of Sheffield. He currently directs the UK New Dynamics of Ageing Programme (www.newdynamics.group.shef.ac.uk) and the European Research Area in Ageing (www.era-age.group.shef.ac.uk). Previously he chaired the European Observatory on Ageing and Older People and directed the UK Growing Older Programme (www.growingolder.group.shef.ac.uk).

Preface and Acknowledgements

This book is the result of a unique comparative research enterprise which brought together experts on key topics in the field of ageing and social policy from their respective bases in Britain and Germany. Through an initial conference, subsequent face-to-face meetings and correspondence the authors of each chapter were able to develop integrated approaches to their topic. Of course in such a complex endeavour, coupled with the time constraints imposed by the initial conference and publication schedule, it was inevitable that our plans would not be realised fully. So, in practice, it was not possible to cement the ideal Anglo/German partnership, within our time frames, for three of the chapters. Two of them, however, are based on extensive research in both countries by their German authors (Chapters 8 and 9) and the third reflects both a close knowledge of the UK system, and unattributed inputs from a leading expert in the UK (Chapter 6). The result is a unique comparative survey of the major policy issues confronting Britain and Germany as their populations age. This survey can be treated as a highly reliable one up to and including 2006. After that date it has not been possible to ensure that every policy change is included.

To embark upon a comparative adventure such as this requires putting trust in many people and a great deal of assistance. From the start our colleagues in the Anglo-German Foundation (AGF) were enthusiastic and supportive. In particular Keith Dobbs and Ray Cunningham deserve our sincere thanks for both their warm encouragement and their hard financial support. The participants in the AGF conference that started the process of creating this book were a stimulating audience who provided much helpful feedback. We have been lucky to work with a talented group of experts in both countries and they responded magnificently to our invitation to work in partnership. Our thanks go to them. The production team in Sheffield, Sarah Counter and Karen Tsui, did some great work. Also special thanks are due to Naina Patel. Finally we want to thank Hazel Woodbridge and Olivia Middleton at Palgrave Macmillan for their patient forbearance during the production of this book.

Every effort has been made to contact all the copyright-holders, but if any have been inadvertently omitted the publishers will be pleased to make the necessary arrangement at the earliest opportunity.

Alan Walker
University of Sheffield

Gerhard Naegele
University of Dortmund

1
Major Policy Challenges of Ageing Societies: Britain and Germany Compared

Alan Walker and Gerhard Naegele

Ageing has been a major focus for social policy in both Germany and the UK for at least three decades and, as the social and economic implications of population ageing have become clearer, this issue has moved steadily up their policy and research agendas. In both countries there are rich traditions of social gerontological and social policy research that, in recent years, have become interlinked both within the two countries and between them. It is partly this existing comparative basis for collaboration between Germany and the UK that started this particular ball rolling. Inevitably too, personal contacts and co-research are critical to such an endeavour, as was the enthusiasm of the Anglo-German Foundation (AGF) and the funding it provided. The latter enabled the in-depth collaboration on which this book is based. In research terms Britain and Germany were ripe for a comparative study of this kind because of their existing research expertise, their similar demographic profile, their common EU membership and their contrasting welfare regimes (Walker, Maltby and Walker, 1997; Naegele and Walker, 2007).

This comparison takes place against the backcloth of the two countries' membership of different welfare regimes. While recognising the limitations of such regime classifications, particularly that all of them tend to overgeneralise and oversimplify, and the fact that the Western European ones have been criticised for being gender-blind and ethnocentric, even the most basic versions do capture key elements of the socio-political, legal and organisational features of different countries' approaches to the provision of welfare (Naegele and Walker, 2007).

The UK belongs to the liberal welfare regime type, which primarily emphasises the roles of both the free market and the family (Esping-Andersen, 1990). The market is considered to be the main arena for

1

the distribution of resources, social security benefits are modest and social rights are rather limited. Consequently, welfare benefits are often means-tested and/or are usually paid at a low level. In contrast, Germany belongs to the corporatist-conservative welfare regime type which was shaped by the Church with a strong emphasis on traditions – such as the family and the pre-existing class and status structure. In this welfare model, social insurance forms the core of the social security system and is thus predominantly linked to a person's occupational status.

Against the background of traditional differences in the respective welfare regimes each country belongs to, this book provides an overview of the ageing and social policy context of the two countries and the key challenges facing them. The issues addressed in this introductory chapter are explored in greater depth in the following nine chapters; then we draw together the various threads in the Conclusion.

Demographic background

Both countries are typically, in demographic terms, part of the oldest region in the world and share its continued ageing. As with other countries in the EU the twin drivers of this demographic revolution are, on the one hand, the decline in the birth rate (in both Germany and the UK to below the 2.1 children per woman necessary for the population to replace itself), which is in Germany more pronounced than in the UK (total fertility rate in 2004 in Germany 1.4, in the UK 1.7), and, on the other hand, the increase in the average life expectancy. Table 1.1 shows that the two countries have very similar old-age profiles currently: 22 per cent aged 60 and over in Germany and 21 per cent in the UK.

Although it is always necessary to enter cautions about population predictions, the future trends suggest the development of a gap between the two countries. By 2050 those aged 60 and over are expected to comprise between 36 and 40 per cent of the total population in Germany compared with 31 per cent in the UK. The main reason for this appears to be the higher projected life expectancy in Germany than in the UK in 2050.

Medium life expectancy (at birth) and further life expectancy (at the age of 60) in Germany and the UK in 2002/04 and 2050 are shown in Table 1.2.

At successively older ages the population of women increasingly exceeds that of men. For example, in the UK currently, women account for 49 per cent of those aged 0–24, 50 per cent of those aged 25–64, 53 per cent of those aged 65–74, 61 per cent of those aged 75–84 and

Table 1.1 Age groups as a proportion of total population in Germany and the UK, 2000–2050 (%)

Age group	2000	2010	2020	2030	2040	2050
Germany						
0–14	15.1	13.3	12.3	12.1	11.3	11.0
15–29	18.0	17.2	15.3	14.0	14.1	13.8
30–44	26.1	20.2	18.8	18.2	16.5	16.2
45–59	19.4	23.0	22.9	18.9	19.7	18.6
60–74	16.4	17.1	18.8	23.2	20.1	20.1
75+	5.0	8.5	10.6	11.7	16.2	17.1
All ages	100.0	100.0	100.0	100.0	100.0	100.0
The UK						
0–14	18.9	16.8	16.2	16.0	15.3	15.3
15–29	19.1	19.6	18.1	16.8	17.2	16.9
30–44	23.0	20.9	19.3	19.5	18.1	18.4
45–59	18.5	20.3	21.3	18.3	19.4	18.6
60–74	13.1	14.9	16.4	18.7	17.1	17.2
75+	7.4	7.6	8.6	10.7	13.0	13.6
All ages	100.0	100.0	100.0	100.0	100.0	100.0

Sources: The UK: UK Government Actuary's mid-2000 based principal projections, 2002; Germany: 11th Coordinated Population Forecast of the Federal Statistic Office, 2006.

Table 1.2 Medium life expectancy (at birth) and further life expectancy (at the age of 60) in Germany and the UK, 2002/04 and 2050

Measure	Germany				The UK			
	2002/04		2050		2002/04		2050	
	Men	Women	Men	Women	Men	Women	Men	Women
Medium life expectancy (at birth)	75.9	81.5	83.5	88.0	76.2	80.7	78.6	83.6
Further life expectancy (at the age of 60)	20.0	24.0	25.3	29.1	20.2	23.4	25.1	27.5

Sources: The UK: UK Government Actuary's mid-2000-based principal projections, 2002; Eurostat Online Databank; Germany: 11th Coordinated Population Forecast of the German Federal Statistic Office, 2006.

72 per cent of those aged 85 and over. In Germany there are compa-rable trends for those aged 0–20 (49 per cent female), for those aged 21–60 (51 per cent female), for those aged 61–70 (52 per cent female), for those aged 71–80 (57 per cent female), and for those aged 80 and over (72 per cent female). The main reason for this discrepancy is that female life expectancy (at birth) exceeds that for men – by 5.6 years in Germany and 4.9 years in the UK. Moreover, although men's life expectancy is projected to improve over the next five decades, so too will women's. Therefore, by 2050, the difference is still expected to be around 4–5 years in both countries (Chapter 4).

Migration has been a significant feature of the demography of both countries and, indeed, has prevented their populations from declining. The UK has longer experience on this front than Germany because of the influx of people from various parts of the Commonwealth start-ing in the 1950s. Among black and ethnic minority groups in the UK the population of older people is currently much smaller than that of the white population – only 6 per cent of black, 7 per cent of Indian, 6 per cent of Chinese and 3 per cent of Pakistani/Bangladeshi per-sons are over the age of 65 compared with 16 per cent of the white group. Progressive ageing of ethnic minority groups is expected in the future (but depends on fertility levels, mortality rates and net immigra-tion). In Germany, older migrants are likely to have the highest and fastest growth rates among older people over the next decades. Due to Germany's past migration policy, among older migrants former guest-workers are over-represented and among them again men living alone. Within the group of older migrants Turkish people are over-represented. However, in contrast to the UK, a special feature can be observed in Germany: in a growing number of cases older migrants tend to com-mute between their home countries and Germany, often spending the summer in Germany and the winter in their home countries. Whereas in the UK social and care services for older migrants are long established, Germany is in the process of adapting existing services to the needs of older migrants (Chapter 6).

Employment and the labour market

Employment is a key determinant of living standards in old age. There-fore whether or not those in their later years of working life (50–65) remain securely in paid work matters a great deal not only to their current but also to their future socio-economic security (Walker, 1980, 1981; Ginn, 2003). In both countries the participation rates of older workers declined steadily in the 1960s and 1970s (Chapter 2). In fact the trend towards early exit from the labour market on the part of older men means that, in both Britain and Germany, barely one-third of men enter retirement directly from full-time employment. The majority reach the initial pension ages in a variety of non-employed statuses such as unemployment, long-term sickness or disability, and early or pre-retirement.

In both countries this trend applies to men rather than women. Among women the cohort effect of increasing proportions of successive generations of women joining the labour force, and their preference for part-time work, has seen either a steady or an increasing rate of employment up to the age of 60. In Germany this applies only to the former West German Länder, which emphasises a major difference between the two countries. The decline in employment rates among older workers has been particularly drastic in the former East German Länder, where the major restructuring that has taken place has led to a collapse in employment opportunities for this age group, among both men and women.

Apart from this special set of circumstances following from German reunification, the main causes of the late-twentieth-century trend towards early exit, in both countries, are a mixture of demand-push and supply-pull factors. Among the more dominant demand-side factors are the economic recessions of the mid-1970s and early 1980s and the higher risks that older workers face in the labour market compared to younger adults, including age discrimination and long-term unemployment. These factors, of course, have a bearing on the supply-side choices that people make concerning early retirement (Walker, 1985; Naegele, 1992, 2004; Taylor and Walker, 1998). So too do public policies and, in both countries, these played important roles in facilitating early exit (pre-retirement in Germany and the Job-release Scheme in the UK).

Considerable research effort has been devoted to the issue of employment among older workers in recent years and, among other things, this shows that people aged 50–65 face widespread age discrimination from employers. For example, in the UK, age restrictions placed in job advertisements are a common barrier to employment. In both countries employers often restrict access to training programmes to workers under the age of 50. Many employers have been shown to hold stereotypical attitudes towards older workers, e.g. in terms of their showing less productivity, more skill deficits, higher rates of absenteeism due to ill health, higher direct and indirect personnel costs, many of these attitudes often based on prejudices that are not supported by research evidence.

Obvious results of both labour market as well as in-company discrimination against older workers are the high rates of long-term unemployment and non-employment among this age group. In both countries, older workers are more likely than younger ones to be unemployed for long periods, and especially for very long periods. They are also much more likely than younger workers to occupy other non-employed statuses such as sick and disabled and early retired.

In view of both shrinking and ageing of the workforces in Germany and the UK, in both countries it is increasingly accepted that activity rates of older workers must be increased and age discrimination must be combatted. In Germany the rising financial costs of declining activity rates of older workers as well as early retirement practice in the last three decades – due to the double-effect of shorter periods of contributing to and longer periods of benefiting from the (PAYG-organised) social security systems – have hastened policy efforts to stop early exit and to keep older workers in employment. In both countries, though more so in the UK than in Germany, there has been a great deal of policy activity in this field in recent years (Chapter 3). This change of paradigm started earlier in the UK than in Germany. Policies in both countries include a range of 'soft' measures such as campaigns as well as financial incentives to persuade employers to recruit, train and retain older workers; and 'hard' ones, such as the implementation of the EU's Equal Treatment in Employment Directive, or – as happened in Germany in 2006 – raising of retirement ages to 67 (beginning step by step in 2013) combined with cutting pensions for those who still exit earlier. Nonetheless, in both countries, there is still a need for continuing actions in the fields of pensions and labour market policies as well as in the fields of in-company age management (Naegele and Walker, 2006) if the employment rate of older workers is to be raised and equal opportunities created.

Income, poverty and wealth

Unlike with the previous topic, this one is a story of contrasting institutions and outcomes. Indeed this topic takes us to the heart of a comparison of the different welfare regimes embedded in each country and their differing impact on socio-economic security in old age. This significance derives from the fact that public pensions provide the cornerstone of Europe's social protection systems and, in most cases, have a strong influence on the nature of the system as a whole. In a nutshell, while the German scheme that was initiated in 1889 followed the Bismarckian path of being earnings-related, the British one, introduced in 1948, took the diverging Beveridgean path of minimal flat-rate provision. As Ginn, Fachinger and Schmähl (Chapter 2) show, the resulting pension systems are different although both of them are based on three pillars. In both countries the incomes of pensioners depend on which combination of the various components of the system they gain access to.

In a comparative EU perspective British pensioners tend to be worse off than those in other Northern countries. The explanation for this lies in the basic structure of pension provision erected in the late 1940s, and the policies of subsequent governments which sought to first enhance (in the 1970s) and then dismantle (in the 1980s) the public pension system. Public pensions were introduced in 1908 and made universal 40 years later, following the recommendations of the Beveridge Report. Unlike Bismarck, Beveridge believed that the best approach to pensions was to create a basic universal floor upon which people could build their own voluntary contributions. This floor has always been set at a low level in the UK (unlike the universal public pensions in Scandinavia) – fluctuating between one-fifth and one-third of male manual earnings. In the 1980s the Thatcher governments subjected this minimal system to substantial cuts, starting first with the decoupling of pension rises from average earnings and then halving the value of the State Earnings Related Pension (SERP), which had been introduced with all-party political support in 1978 and was due to reach full maturity in 1998. As well as cutting public pensions, positive inducements were offered for people to opt out of the state sector and into the private one. The emphasis of the Blair and Brown governments to date has been on means-tested social assistance rather than raising the basic pension, although there are plans to do this (Chapter 2).

Despite the fact that social security is by far the largest public expenditure programme and older people receive over half of it, there is still an association between low income and poverty and old age. There is no 'official' poverty line in the UK. Instead a widely used definition of poverty is based on 60 per cent of average income before housing costs. On this basis, the latest figures show that 20 per cent of pensioner couples and 27 per cent of single pensioners were in poverty in 2006/07, over 2.5 million people. At the same time there is no doubt that the income of some pensioners has increased substantially, largely due to the impact of occupational pensions. There is, in fact, a polarisation of older people between poverty and affluence. The income of the top 20 per cent of pensioners has increased by 70 per cent since 1979, whereas that of the lowest 20 per cent has risen by only 38 per cent.

The poorest pensioners tend to be very elderly women while the most affluent tend to be younger couples. Black and ethnic minority older people are more likely to be poor in old age than their white counterparts. These differences signal the failure of the British pension system to provide for those in low-paid part-time or discontinuous employment, especially women, and this represents a huge policy challenge. At the same time UK pensioners are reluctant to subject themselves to means-tests and, consequently, there are at least 500,000 living below the Pensions Credit level. Public pension spending is lower than the EU average (11.5 per cent of GDP compared with the average of 12.7 per cent and the German rate of 13 per cent) and, therefore, there has not been an economic imperative to reduce it (the imperative in the 1980s was ideological). However, there are current fears among experts in the UK that the existing structure of pension provision is simply not adequate to guarantee the income of future pensioners. These fears are based on three factors: the low level of the basic National Insurance (NI) pension; stock market volatility and the risk this creates for those with private defined contribution pensions; and the recent collapse of the occupational pension system that had previously guaranteed many older people a decent standard of living. Public pension provision was undermined in the 1980s and it is questionable whether the private market is able to prevent the risk of poverty in old age in the face of uncertainty in world stock markets and the changing nature of employment and careers (the 'destandardisation' of working life, see below). Thus the main pension policy challenge in the UK is less about long-term financial stability, as in some other EU countries, and more about how to ensure an adequate standard of living for tomorrow's pensioners and, most urgently, how to eradicate poverty among today's.

The financial situation of pensioners in Germany contrasts sharply with that of their British counterparts. Of course in any pension system that is geared to employment income the inequalities forged there will be mirrored in retirement. So, those with high incomes and continuous employment have the highest pensions (as in the UK) whereas those with low incomes in employment and/or short or discontinuous working lives run the risk of experiencing poverty in old age (also as in the UK). Nonetheless, while poverty in old age was a widespread phenomenon in Germany until the 1970s, only a minority of older people are poor today. Evidence indicates a high level of satisfaction of seniors with their financial situation. At present a state of wealth in old age is more common than one of poverty. This can, for example, be gleaned from the increased income from assets of seniors but also from the financial transfers from the generation of the elderly to the generations of their children and grandchildren.

Despite a high level of financial satisfaction on the whole there are significant differences in the amount of benefits received: people with a low average income during their working lives and short/discontinuous periods of employment tend to be inadequately provided for financially in old age. This applies, for instance, to many women who have raised children or have cared for relatives but also to many unqualified male blue-collar workers. This is also true for many older migrants who have low-paid jobs and/or have, on average, only attained significantly shorter periods of insurance in the statutory pension insurance scheme for all employees. It is also partly true for older people in the new *Länder* who, on average, have a lower income from private assets and from private company (occupational) pension schemes and/or who were often compulsorily sent into early retirement and thus had a shorter working life (and insurance).

In the future those employees who do not have long and continuous working lives will be at risk. This group includes many long-term unemployed people, women with career interruptions (e.g. for child and/or elder care), or those (many of them also women) with discontinuities in their working lives and/or flexible working-time arrangements. The latter is an expression of the growing trend of the 'destandardisation' of working. It is also partly due to the consequences of social change (e.g. the increase in the rate of divorce). Particular financial and income risks also await those who cannot (or do not want to) pay into private insurance schemes introduced on a voluntary basis in 2002. Further future income risks are to be expected for those who do not fulfil the preconditions on reaching the new retirement age of 67 and therefore have to accept actuarial cuts (Chapter 2). Thus, in the future, there will be increases both in old-age incomes for some and, for others, in poverty in old age.

In sum, public pension policy in Germany faces two challenges. First, against the background of still-high unemployment, low activity rates of older workers and demographic changes, it is important to secure the long-term stability of pension schemes. This is equally true for systems based on the pay-as-you-go (PAYG) principle and private pension schemes based on the capital stock system. Apart from the decreasing number of contributors, the growing number of benefit recipients on the one hand and the higher life expectancy on the other hand influence the financial solidity of the pension schemes. Second, in Germany there are an increasing number of discontinuous and/or unstable working arrangements (part-time work, temporary work contracts, decisions in favour of financially less secure self-employment, jobs not subject to compulsory insurance, etc.). These have become the only forms of employment open to many. The consequences of unemployment, which may sometimes be long-term, and compulsory early retirement, must be added to these risk factors.

Health and health care

Good health figures prominently in all subjective assessments of quality of life in old age and it is related to age: higher proportions of successively older age groups report poor health and low levels of functional capacity. Older people are more likely than younger people to experience several co-morbidities. Therefore the projected increase in the numbers of very old people in coming decades presents a potential challenge to the health care systems of both countries (Chapter 8).

A longstanding observation of research in both countries is the paradox of gender differences in health in old age: men are more likely to die at earlier ages than women, but women suffer higher levels of morbidity. As noted previously, older women are prone to poverty and this socio-economic disadvantage is associated with ill health. Older women are much more likely than men to experience restrictions in mobility, ability to perform household tasks and self-care. There are also strong associations between a person's last main occupation and health in later life. For example, among those aged over 80, professional men report 20 per cent more 'good' health than other class groups. Older people with the highest incomes report the best health. In both countries older migrants have a generally higher morbidity risk and a lower life expectancy.

In terms of the future direction of health status in old age the so-called bi-modal concept represents the current state of knowledge. It is predicted that the succeeding age cohorts will have lower rates of morbidity – due to medical progress, better health and higher health awareness. At the same time, however, due to growing further life expectancy and the strong increase in the number of very old people, the number of the chronically ill will also grow. The reasons for this are the new and in many cases irreversible illnesses like dementia, which occur more often in late old age. In other words: the health status of succeeding cohorts is improving in general but, at the same time, the proportion of chronically ill older people will also increase.

In both countries, cognitive impairment is by far the most significant reason why older people enter long-term institutional care. With regard to the mental health of the whole older population, they are two or three times more likely to be depressed than to have dementia. Depression is not part of normal ageing or age-related, but it is more common among older people who suffer from major health conditions such as Parkinson's, stroke, neurological disorders or dementia. Without treatment depression can become a chronic disorder. In later life depression is a largely untreated and undetected condition. People with low coping resources, such as those in poverty, are more likely to report depression. Very little research has been done in either country on older people in migrant groups, who might have fewer resources to cope with mental health problems (Chapter 6).

When it comes to the future challenges facing their health care systems, although population ageing is common to both Britain and Germany, and in similar measure, the nature of their existing health care systems implies a different scale of challenges in the two countries. The UK's National Health Service (NHS) has a long-term legacy of underfunding and, therefore, undercapacity in key areas, such as dementia care and stroke rehabilitation. In addition, age discrimination has been identified as a core problem in the NHS (this form of rationing does not exist in German health care). Underfunding of the NHS is one reason why the future cost of health care for an ageing population has not been such a high-profile issue in the UK as it has in Germany. In both countries, though, age-related health expenditure per capita has been rising consistently and, against the background of demographic ageing, the question of how to finance future medical treatment for an overall ageing population characterised by multi-morbidity and chronic diseases will undoubtedly be of central concern to policy makers in both countries. Important contributions to the balancing of needs and finances, in both countries, will be policies to 'join up' the health and social care services, to enlarge and to improve the possibilities of domestic care for those elderly with chronic diseases, a stronger emphasis on prevention and the promotion of healthy ageing across the life course (Walker, 2002a, 2005).

Nursing and social care

If older people need care, in both countries it is most often provided by themselves or by close relatives. If the need is substantial and/or there are no available relatives then they may be cared for in residential or nursing homes, depending on the level of need (Chapter 9). In both countries the need for personal care rises with age and is of particu-lar significance for the very old. However, the need for care, especially nursing care, does not depend so much on the actual age but more on the remaining life expectancy.

Institutionalisation is a feature mainly of late old age in both coun-tries: for example, in the UK only 1 per cent of people aged 65–74 live in care homes or hospitals but this rises to 25 per cent for those over the age of 85. Thus, for a significant proportion of older people such insti-tutions provide end-of-life care. Among those older than 80 living in long-stay institutions in the UK, women outnumber men by 2.3 times. In Germany around 30 per cent of older people who need personal care are resident in nursing homes or other institutions. However, as in the UK, the proportion rises with increasing age. Thus the vast majority of older people needing care and support, even in advanced old age, receive it in their own homes or those of close relatives. The main family car-ers of older people are usually either adult children (mostly daughters) and/or spouses. Thus a great amount of care of older people is provided by other older people (in the 50–60 and 80+ age ranges).

Public provision for social care reflects the different welfare regimes identified earlier as broadly Bismarckian and Beveridgean. In Germany the strong family-orientated tradition and federal governmental struc-ture meant that care was largely a personal and local matter, with NGOs playing important roles as providers. Then, in 1994/95, after a long dis-cussion lasting more than twenty years, the German government took the radical step of introducing a national long-term care insurance pro-gramme. In effect this added long-term care risks to the Bismarckian insurance system. In contrast the UK has followed the Beveridge path in social care as well as pensions. This means that social care is provided as a safety net and, increasingly over the past two decades, on a means-tested basis. Nursing, on the other hand, including community nursing, is part of the NHS and therefore free of charge. Needless to say, in the UK the issue of cost-shunting between health and social care has been a major one for decades. Because social care services are the responsibil-ity of local authorities (as providers or commissioners) their levels and quality vary between areas, as do the fees charged. In most urban local-ities older people would expect to find home care / home help services, day care and meals-on-wheels provision supplied mainly by private and voluntary providers. Services are more patchy in rural areas.

In Germany social services for older people are often subsumed under the terms 'open work with seniors' or 'open promotion of seniors' welfare'. Traditionally the local authority districts are responsible for organising these services which, according to the subsidiarity principle, cooperate closely with non-state charitable institutions and other privately organised welfare associations. Offering open services for older people is one of the essential communal tasks in the field of promoting welfare, which is guaranteed by constitutional law. For the future this is particularly true for those kinds of services that support the rapidly growing number of families with older members suffering from dementia. In this context new types of housing for older persons suffering from dementia are called for in Germany, based on the idea of independent living in so-called 'group-homes'.

However, these services are not available everywhere in Germany to the same extent. Lower levels of services are found particularly in rural areas, areas lacking infrastructure and/or economically weak areas. In addition, interlinkage between the various services tends to be an exception. Apart from their financing, which very often does not have a secure basis, other problems arise because these services often only reach certain subgroups of older people. For instance, older people belonging to relatively deprived groups (e.g. older migrants, those living on their own) benefit less than the average from such services. There is thus a need for action to improve the financing of these services, broaden their base, their coordination and interlinkage, to develop quality standards hitherto almost absent, and a more exact orientation towards all target groups as well as a need for more user involvement and user participation. Problems of access will, however, remain because these services are sponsored by different organisations.

Looking to the future it is predicted that, despite a long-term trend towards reductions in morbidity, the need for care will rise. This is due to the combination of increasing life expectancy and chronic conditions dementia. On the other hand there are changes in family structure, resulting from falling fertility and rising marriage breakdown, which will reduce the pool of potential family carers. In addition the growth in labour market participation among women places particular strains on the main carers of older people. To avoid an increase in the hospitalisation rate of older people in need of care in Germany, recently the question of how to provide for more and at the same time qualified professional carers in the future has entered the political agenda. One of the answers given by many experts (and already in practice in a rising number of families affected) is to attract foreign professional carers to work in Germany.

Housing

The vast majority of older people in both countries live independently in private dwellings: less than 6 per cent in Germany and 7 per cent in the UK live in residential or nursing homes. Also, in both countries, the majority of older people own their dwellings. Choosing where to live is dependent on many factors in an older person's life situation. Often the decision about a change in living situation is made under stress of having the present arrangement break down and, to complicate the decision process, there is the added pressure of an unpredictable future in terms of functional ability and state of health. Most older people do not want to move to a residential or nursing home, because of their love of their own home and the negative views they have of residential life. The sense of self-identity is altered by a move to a residential home, which is often seen as 'living in someone else's home'. The key element in influencing people's subjective well-being is the extent to which they control their decisions and manage their activities of daily living. Frail older people often have others making major decisions for them, including when they have to move to a nursing home or residential home.

In both Germany and the UK most housing stock is not ideal for growing older in, because it is unsuitable for adaptation, or houses might be too big, expensive to run, in the wrong place or otherwise inconvenient. Some older people feel constrained by their housing environment and see it as completely negative and disabling ('architectural disability'). What older people want is suitable mainstream housing, designed with changing needs in mind. In the future, when older people will become the majority, the emphasis in design might shift towards their needs as part of the mainstream and not as special-needs provision. This might give all older people, and not only the better-off, the right to choose their living environment and lifestyle rather than being forced into a narrow band of options depending on health status or temperament.

Sheltered housing has been developed in both countries as a resource for older people, meeting their housing and support needs as well as being an alternative to residential care, which is expensive and not always needed by people unable to stay in their own homes because these were too expensive to run and repair, or were situated in areas of high crime or far away from supporting relatives who would help if needed. Sheltered housing offers a sense of security and contact with other people. In these communal environments older people find it easy to share role models and gain reference groups, where personal self-evaluation or resolve is strengthened by the age-segregated environment.

In Germany various new forms of housing have been experimented with successfully, such as those suffering from dementia sharing an apartment or house. New forms of housing are also being tested for disabled older people. For those who are healthy, there are so-called 'alternative' living arrangements like sharing an apartment or flat or multi-generational apartments, but these are very rare, usually still in the phase of testing, and do not attract more than very small minorities. All special housing arrangements reflect attempts to react positively to the heterogeneity of the older age group.

In both Britain and Germany there is general agreement that the promotion of living in normal housing should have priority, and this is overwhelmingly what older people want. Nonetheless, if heterogeneity and pluralism are in demand, which is even more likely for future cohorts of older people, then these special forms of housing should also be developed and promoted. Regardless of what special provision is made the normal housing stock needs attention. Here it is particularly important to adjust houses and apartments to the changes in physical constitution typical for old age. Housing counselling, the adjustment of living space, modernisation measures meeting the new demands, especially as far as barrier-free living space and access to buildings and apartments are concerned, as well as the refashioning of bathrooms / sanitary areas to meet the needs of older people, are important measures. The use of the new information and communications technologies (ICTs) is also part of the current attempt to promote autonomous forms of living for older people; however, this has to battle against both insecure financing as well as a widespread tendency of refusal among the very old.

Social networks and social support

Social networks are widely regarded as critical to the quality of life of older people, but there is very little research evidence to show just how important they are and what contributions they make (Chapter 5). What evidence there is tends to focus on the role of families. There are increasing proportions of families spanning three or more generations. In the UK close to three-quarters of the population, with the exception of those in their fifties, are part of three-generational families. Many mothers with children younger than 18 receive help from their own mothers, and half of mothers aged 50 and over receive help from their eldest child. Because they live longer, women are more likely than men to be involved with family members across several generations.

Social support consists of what the social network members or supportive others actually do for the individual. Although the types of support people receive from their network are quite varied, most of them fall within two general types: instrumental and emotional support. Instrumental support refers to aid that helps individuals deal with life demands (e.g. lending money or cleaning the house, caregiving). Emotional support pertains to actions, gestures and words that engage an individual's thoughts, feelings and perceptions (such as making them feel loved or liked; or affirmation, communicating that we are in agreement with the individual's perspective on a problem). Also, in both countries, many members of the wealthy grandparent generations support their children and their grandchildren financially in various ways.

The support networks of older people in Germany and the UK are constituted predominantly by close family members such as spouse, children, grandchildren or sisters and brothers, followed by other relatives, friends, neighbours and acquaintances. However, network composition is affected by age, marital status, feeling of family closeness, whether or not the individual has children, and health status – to name only a few important variables. For example, for older people who are in need of care and/or live in an institution, professionals can also become important network members, providing not only instrumental but also emotional support.

It is argued that reciprocity – the perception of an equal exchange of support – is a basic aspect of normal well-functioning support networks (Qureshi and Walker, 1989). Both British and German research shows that there is a history of support exchange among family members over the life course, so that intergenerational reciprocity within families is one dimension of solidarity. With regard to non-kin support, however, short-term reciprocity is a significant precondition for the provision of social support. In addition, family links have a strong obligatory component, whereas friendship is optional.

The social network of an older person is threatened and weakened in many ways. First, the loss of important network members and support providers through death (e.g. a spouse). Second, older people themselves might not be able any longer to 'invest' in their network in terms of reciprocity because they are too ill and fragile. The loss of significant support providers might be the consequence. Therefore, older people who do not have children or other close family members who could give them instrumental and emotional support if they need it are in danger of not having a reliable social network. The same is true for older people living in nursing homes. Furthermore, in old age, social networks very seldom are completely newly formed but, rather, are likely to be the expression of a life-long development. In this vein, the concept of a 'convoy' of social relations is useful. It assumes both an intra-individual and an inter-individual perspective on social relationships, which takes into account the continuities and discontinuities, the stability and change, the growth and development of both the individual and those with whom he or she interacts.

As a result of socio-demographic changes, there are further factors which influence the size and composition of social networks and the provision of social support. First, it is doubtful whether the family can meet the need for help and care of its older members over the long term. As a result of women's growing participation in education and employment since the 1960s, and the increasing tendency for a pluralisation of lifestyles and the trend towards individualisation, women have acquired more scope for action and more options for organising their lives. Consequently, there has also been an increase in ways of life not involving marriage and a family. This trend finds expression in rising divorce rates, even after many years of marriage, a growing number of one-person households, and an increasing proportion of women who remain childless throughout their lives. At the same time, second, family structures are becoming ever more complex and diversified. German terms which translate as 'chain marriage and divorce', 'serial marriages', 'multi-parent families' and 'patchwork families' attempt to conceptualise these new family structures. In the long term, this process of diversification may mean that family ties will not imply the same strong element of obligation as they do now. In this context, it is argued that family ties are no longer so much a matter of course; they are becoming weaker and more fragile, more dependent on personal involvement and external circumstances. Such speculations of course require careful research. Lastly, labour market policy constraints and the associated need for job mobility and flexibility are increasingly forcing families to become widely scattered. However, this imposes a considerable restriction on the opportunities to support relatives in need of assistance and care.

Also, in a wider perspective, the question remains whether the demographic changes will have an influence on the intergenerational relationships in society as a whole, bearing in mind that the cohesion of a society is determined by the solidarity of its members. With regard to 'generational conflict' it has to be realised, first, that older persons are not only the recipients of social support; within the family they also provide it to a large extent. For example, grandparents can play important social support roles in assisting working mothers with child care. There is evidence too of increased support from grandchildren to older people following widowhood. Second, to avoid any resentment, policy measures can help to improve (non-kin) intergenerational relationships systematically in all life-domains (e.g. in housing or at the workplace). This means, on the one hand, that older people have to be encouraged to be more 'active', i.e. one has to make them – and also younger persons – aware of the fact that old age and having a lot of competencies are not incompatible. Third, society has to give older people different opportunities to practise 'active ageing' in everyday life. For example, a society can facilitate voluntary work by older people with younger ones, for example by providing financial incentives.

Social and political participation

To guarantee the social and political participation of older people is another central task for an ageing society (Chapter 10). In this context, one of the important questions refers to how older people themselves can be actively engaged in creating and influencing their living conditions and their environment. This not only affects their participation at the micro level (e.g. private and institutional housing) but also at the macro level (e.g. political parties and national or local governments). Also, older people have to be involved in the process of 'quality assurance' regarding the care services and institutions which are mainly designed for them.

Social and political participation have been increasing among older people in the UK as a result of the improvements in health and living standards in this age group. It is only very recently that this has become of interest to policy makers as a result of their commitments to active and healthy ageing and greater participation in the policy process. Voluntary activity is common in the UK and some two out of five volunteers are over the age of 60. The most likely people to volunteer in Germany as well as the UK are those from a middle-class occupational background. Older people share many leisure pursuits in common with younger people but societal ageing has also brought with it increasing amounts of specialist provision for older people, for example reminiscence, open learning, intergenerational activities and IT training. These facilities are usually provided by voluntary sector organisations such as Age Concern. Local authority adult education services usually provide reduced rates for pensioners, although, perversely, these services have been cut back heavily in recent years. There are many local authority special programmes in leisure centres, sporting facilities and swimming pools.

The UK has also seen the rise of political participation among older people in the last decade. This has been a grass-roots movement that has responded to policy rather than (until very recently) being encouraged by it, the main impetus being cuts in pensions in the 1980s and 1990s. Numerous local pensioner action groups and pensioner forums have sprung up independently from the established voluntary groups working on behalf of older people. Following the 1997 election of the Labour government, encouragement was given to the participation of older people in local government by the Better Government for Older People initiative. In practice, however, the real involvement of older people in decision making is very limited.

In contrast to Britain, German research points to the fact that the majority of older persons withdraw from social and political activities. There is hardly any grass-roots movement and 'senior power' is an artefact of official 'encouragement'. Although the 'disengagement theory' is still quoted in this context, it is not yet known whether this development is a real 'age' effect or rather a cohort effect. Thus, the question of an appropriate form of representation arises. Special attention has to be given in Germany to the political participation of older persons at the local, regional and national levels.

Another key aspect of participation concerns the health and social services and, specifically, the issue of who should decide on the treatment or care to be provided. In the past this was not an issue: it was the task of professionals to decide on the appropriate care or treatment. Over the last decade, however, pressure has built up among various groups of service users for more involvement in such decisions. In the UK this 'user involvement' movement was conceived by disabled people but, recently, older people and their carers have played key roles too. The policy process has responded gradually, first recognising the rights of carers in 1996, then the rights of users to consultation and, most recently, patients' and users' forums have been built into the health and social services.

In Germany, user involvement of older persons with regard to social services is underdeveloped. The wishes and needs of older people themselves are hardly taken into consideration when services are offered and moreover, so far, they are not involved when it comes to quality assurance and quality control. Very often it is assumed that older persons who need support and care cannot take on the role of a 'critical' consumer. However, in the wake of the growing privatisation of social services which are then subject to the laws of the market and competition, effective concepts of consumer protection are needed. A high priority in this area has to be protection against the abuse of vulnerable older people.

A further important question which is being discussed in Germany recently is how the (often unused) potential of seniors might be better used by society. This question is embedded in the overarching discussion of how to arrange the intergenerational contract in a more balanced way (Chapter 7). The popular slogan of an 'active ageing' is increasingly regarded less as an individual challenge and more as a societal task which has to be taken on by those older people who have no other obligations (e.g. elder care, support of single-parent families, voluntary activities). Recently, the federal government launched a programme called 'Active in Old Age', based on the EU model of 'active ageing', which particularly aims at a stronger participation of older persons in problem solving concerning societal and political issues, particularly at the local level. This new programme is based on the fifth federal report, issued in 2006, on the situation of older people in Germany under the title *The Potentials of an Ageing Society*, which suggested an increase in the active contribution of the older population in order to contribute to generational equity, productive ageing and, reading between the lines, 'to pay back what society is spending on them'.

In this context it is also argued that 'active ageing' in future should have a different meaning compared to the hitherto prevailing interpretation, which primarily refers to individual preferences and attitudes. Are the 'new old' entitled to societal inactivity, or is there a duty in terms of being socially obliged to the world around them, for example as active participants in the so-called 'silver economy' (Chapter 4)? In principle, this concept of 'active ageing' might harmonise with a widespread positive attitude among early retirees towards (meaningful) activities in retirement (Walker, 2002a). It might also equate, however, with the prevailing 'public burden scenario' of population ageing – recently accompanied by a public discourse about the 'fertility crisis'. As in the UK, reducing age-related expenditures is a political goal particularly of neo-liberal scientists and policy makers.

So far, the 'new politics of age' in Germany only refers to old-age income security systems (Chapter 2) and recently to rising medical expenditures (Chapter 8). However, as is feared by social policy experts in the wake of a further demographic ageing of the population, this might change soon and be transferred to different policy fields (Chapter 10). In other words, under these circumstances, on the one hand the new rhetoric of 'active' and 'productive' ageing can be seen as an attempt to counteract the prevailing 'public burden scenario' by explicitly drawing attention to the competencies and potential of older people. On the other hand, however, it could also be interpreted as preparing ground for a policy continuing or even reinforcing the restriction of expenditures on population ageing. Thus an ageing population, rhetorically and ideologically promoted as 'active', 'productive' and 'full of hidden potential' is more likely to accept financial sacrifices, such as cuts in public expenditure or privatisation, and thus itself to contribute to the containment of the costs of an ageing society.

2

Pension Reform and the Socio-economic Status of Older People

Jay Ginn, Uwe Fachinger and Winfried Schmähl

Introduction

Pension privatisation – shifting the balance of pension provision away from public provision and towards the private sector (occupational and personal pensions) – has varied across Europe, depending on dominant societal values and political feasibility. Reforms have been justified by reference to demographic change. It has been claimed that state pensions based on the generational solidarity of a pay-as-you-go (PAYG) system will be unsustainable, placing an unfair burden on younger generations and provoking conflict between workers and pensioners (Johnson *et al.*, 1989; World Bank, 1994). Privately funded pensions, it is argued, would avoid this and should therefore be promoted while state pensions are reduced.

Many EU countries, including Germany, have followed Britain in promoting a shift towards increased private pension provision. In this chapter, we compare these two pension regimes in terms of their history and structure; trends and policy debates; the effects of policies on pensioners and workers, and challenges for the next ten years.

Historical background

Origins of public pensions

In Britain, the model of social security, laid down in the Beveridge Report (1942), was designed to provide a minimum income in old age while giving maximum encouragement to voluntary saving through private pensions and other forms of saving. The National Insurance (NI) pension scheme, funded by employers and employees, soon became

PAYG. This was in order to allow pensions to be paid immediately to older people. Beveridge's 'subsistence' amounts of pension were not based on any objective study of need and for most of the postwar period have been below the level of means-tested benefits and thus inadequate to live on. Between 20 and 30 per cent of pensioners required means-tested supplements (Dilnot *et al.*, 1984). The flat-rate pension replaced only 17 per cent of male earnings for most of the period from 1948 to 1971, but rose to 20 per cent in 1981. Women were ill-served by the Beveridge scheme, whose design assumed that married women had no need of pensions in their own right. However, women's state pension age was 60, compared with 65 for men.

Several basic features of the German system of old-age protection date back to the late nineteenth century, when there was discussion on capital funding versus PAYG as well as on earnings-related versus flat-rate public pensions. Instead of a tax-financed flat-rate pension (which Bismarck originally planned) the political decision was to use an insurance model, although with a weak relationship between earnings and pension and low pension benefits. In 1889 the statutory (social) pension insurance was established as the third branch of social insurance. Financing was mainly from employers' and employees' (earnings-related) contributions but also from general tax revenue. The major political objective was to avoid poverty in old age and reduce the burden on local authorities of poverty relief. As in Britain, financing of the public pension scheme during the next decades was mostly on a PAYG basis, initially to enable payment of benefits right from the start of the new scheme. Later, it was continued because of a loss of accumulated capital funds during periods of economic crises and inflation.

Growth of second-tier pensions in the 1960s–1980s

From 1961, a small state graduated pension was introduced in Britain to supplement the basic pension. Occupational pension schemes were only available to a small minority of the workforce and this reform extended an earnings-related scheme to all workers. However, it was too little and too late to stem the demand for private provision (Hannah, 1986). When the Graduated Pension was introduced, occupational pensions meeting certain standards were allowed to 'contract out', so that part of the National Insurance contributions of both employer and employee were paid into a funded occupational pension scheme. Most such schemes provided a defined benefit (DB) pension at retirement, based on the individual's years of pension scheme membership and final salary. Occupational pension coverage rapidly expanded from only 13 per cent of the workforce in 1936 to 47 per cent in 1967, falling slightly to 46 per cent in the 1970s (Hannah, 1986). Beveridge's scheme had 'furthered the conditions within social security for the growth of a multi-billion-pound enterprise of private pensions' (Shragge, 1984, p. 33). British employers' welfare payments, as a proportion of total employee remuneration, doubled between the mid-1960s and the 1980s (Green *et al.*, 1984).

In contrast, the policy emphasis in West Germany in this period was on making the public pension scheme more generous, with a paradigmatic reform in 1957. Dynamic elements were added, in particular aligning pensions to the development of earnings (Schmähl, 1999, 2004). A major objective of this reform was that public pensions should not only prevent old-age poverty but should replace the individual's earnings to a certain degree. The dynamic element meant that pensioners could now share in the fruits of economic development. This radical reform was achieved despite resistance from employers and other industrial organisations, as well as the Federal Reserve Bank (Bundesbank). They feared that a dynamic pension would increase the inflation rate and undermine the ability and willingness to save for old age in private-capital-funded form. These fears, however, proved to be wrong, as occupational DB pensions continued to develop, supplementing instead of replacing contributions to the public scheme (Schmähl, 1997). Occupational pensions were mainly confined to workers employed in large companies.

In Britain, radical improvements in public pensions did not occur until 1975. Under a Labour government, the basic pension was formally indexed to rises in national earnings or prices, whichever was the higher, and the pension needs of women were addressed by Home Responsibilities Protection (HRP) which allowed years of family caring to count towards entitlement. A new State Earnings Related Pension Scheme (SERPS), based on the best 20 years of earnings, replaced the meagre Graduated Pension. The new state pension package, if allowed to mature, would have provided a more adequate pension income, redistributing towards the low paid and minimising the adverse effect of women's caring responsibilities. However, these changes were overtaken by the neo-liberal reforms of the 1980s.

In West Germany too, the generosity of the public pension scheme was further increased in the early 1970s, among other things by introducing an early retirement option at 63 for men, instead of the normal retirement age of 65, without actuarial deductions from the full pension. For women, 60 had been the normal retirement age since 1957. Labour force participation rates of older males declined steeply (for those aged 63 from 67 per cent in 1972 to about 20 per cent in less than twenty years, a steeper decline than in Britain. Together with other politically decided benefits and the longer duration of receiving benefits because of increased life expectancy, this increased the costs of the pension scheme. The combined contribution rate increased from 14 per cent (1957) to 18 per cent (1973).

Retrenchment, 1980s and 1990s

The British Conservative government elected in 1979 brought an individualistic, competitive ideology, expressed in privatisation of many aspects of welfare. Reforms included cutting state pensions and promoting new personal pensions as an alternative to SERPS. Since 1980, the basic state pension has been indexed only to prices, eroding its relative value from 20 per cent of average male earnings in 1981 to around 15 per cent in 2004. SERPS was scaled back by 1986 legislation, basing the pension on average earnings over 49 years (men) and 44 years (women). In 1995, legislation brought further cuts in SERPS and increased women's state pension age from 60 to 65, phasing in the change from 2010 until 2020. The SERPS widow's pension was to be cut from 100 per cent to 50 per cent from 2002.

Personal pensions are individual portable defined contribution (DC) accounts whose fund must be annuitised at or during retirement. To encourage workers to switch to a personal pension, a financial incentive was paid, taken from the National Insurance Fund. This, with high-pressure sales techniques, persuaded many employees to leave SERPS, or a defined benefit occupational pension scheme, for a personal pension. Personal pensions have many drawbacks. First, investment risk is individualised instead of being shared among a pool of workers. Second, charges for administration, investment management and annuitisation can reduce the value of contributions by 45 per cent (Murthi *et al.*, 2001). Flat-rate, front-loaded fees were disproportionately high for the low paid and those with breaks in contributions. For an estimated 30–40 per cent of personal pension account holders, charges exceeded the amount contributed (Disney and Johnson, 1997). By 1993, 5 million individuals had opted for personal pensions, many finding themselves worse off than if they had remained in their occupational pension or SERPS – the mis-selling scandal (Ward, 1996). As an alternative to personal pensions, some companies joined group DC pension schemes, reducing administration costs and sharing investment risk.

The Conservatives' pension policy was broadly maintained by the 'New Labour' government from 1997; an aim was to increase private pension provision from 40 to 60 per cent of the total (Department of Social Security, 1998). Reforms in Britain from 1975 onwards are summarised in Table 2.1.

In West Germany too, the improvements in the public pension scheme were reconsidered a few years later, in 1989. The optimistic economic expectations that stimulated the 1972 reform had changed (the first oil price shock), resulting in several ad hoc measures to reduce rising pension expenditure. Demographic developments (reduced fertility and increasing life expectancy) together with changing labour market conditions were predicted to increase the ratio of pensioners to contributors in the pension scheme. The expected higher contribution rates stimulated discussion on a further major pension reform.

Table 2.1   British pension reforms since 1975

Old Labour, 1975–79

Basic state pension indexed to average earnings (or prices if higher)
Home Responsibilities Protection (HRP) introduced for child care and elder care
State Earnings Related Pension Scheme (SERPS) introduced; 'best 20 years' formula
SERPS widow's pension set at 100 per cent of deceased husband's pension

---

Conservatives, 1980–96

Basic pension indexed only to prices
Legislation to raise women's pension age from 60 to 65 (to be phased in 2010–20)
SERPS accrual rate cut
SERPS based on average lifetime earnings (provision for HRP from 1998)
Legislation to halve SERPS widow's pension (from 2002)
Personal pensions promoted with financial incentives (available from 1988)

New Labour, 1997–

Basic pension remains price-indexed
Stakeholder pensions – personal pensions with limited charges (2001)
HRP in SERPS not introduced in 1998 as allowed for
SERPS replaced by State Second Pension (S2P) (2002)
Means–tested benefits indexed to average earnings; new Pension Credit taper (2003)
Financial Assistance Scheme to compensate for failed occupational pensions (2004)
Pension Protection Fund set up, financed by levy on occupational schemes (2005)

This was decided in Parliament on November 9, 1989 (the day on which, in the evening, the Berlin Wall was unexpectedly opened). Most elements of the reform package were to be implemented in 1992 (see Appendix to this chapter).

The 1992 Pension Reform Act aimed to slow down rising pension expenditure and the consequential rise in contribution rate necessary to balance the budget. The core of the reform package consisted of several interacting measures intended to establish a 'self-regulating mechanism' (Schmähl, 1993). Pension benefits were to be indexed to net instead of gross average earnings. Whereas net pensions had increased more than net average earnings in preceding decades, these would in future increase in parallel, resulting in a stable net pension level of 70 per cent for pensioners with 45 pension points (earned, for example, through 45 years on average earnings). A federal grant to cover pension benefits aiming at interpersonal redistribution remained linked to average gross earnings (as before) and to rises in the social insurance contribution rate. If an increased contribution rate became necessary, the pension-indexing rate would be reduced and the federal grant increased, to limit the required rise in contribution rate. The 1992 Reform Act clearly characterises the social pension insurance as a defined benefit scheme with a distributive goal: to stabilise the net pension level. It was expected that up to the year 2030 the contribution rate would increase to about 27 per cent, instead of over 36 per cent without the reform measures. In addition, from 2001, a reduction of 3.6 per cent of pension for each year of early retirement (before age 65) was introduced gradually in order to encourage later retirement.

Although the long-term perspective on financing German social insurance changed little after 1989, a new intensified pension debate concerning the projected rise in the contribution rate started in 1996. Employers, other industrial organisations and many politicians demanded that the combined contribution rate should not exceed 20 per cent. Non-wage labour costs, especially employer contributions, became a decisive factor in the debate on international competitiveness of German firms. The contribution rate increased from 19.2 per cent in 1996 to 20.3 per cent in 1997 and 1998, although about 2 percentage points of the contribution rate were due to the costs of German unification and another percentage point resulted from early retirement measures agreed among trade unions, employers and politicians to use social pension insurance to facilitate replacing older workers with younger ones, thus reducing unemployment. The contribution rate was expected to increase to about 26 per cent in

2030, due to population ageing, which would place a PAYG scheme under strain. In 1997 another cost-cutting reform was proposed by the Christian Democratic–Liberal coalition (see Schmähl, 1999) but this was not implemented by the Social Democratic–Green coalition formed in 1998.

The new government in 1998 declared that social benefits should be focused more on those who need them and decided on a paradigm shift in pension policy: to replace part of the PAYG social pension insurance with privately funded pensions. This was seen as the only way out of perceived difficulties. Debate was framed by concepts like fiscal sustainability (due to a looming demographic crisis in PAYG schemes), intergenerational equity (balancing the incomes of pensioners against a rising burden for younger cohorts) and the challenge of globalisation (requiring lower employer contributions to maintain competitiveness). Partial pension privatisation was intended to ensure fiscal sustainability of pensions and make all present and future pensioners better off, an outcome seemingly confirmed by the booming stock markets in 1998. Dissenting views – initially from left-wing MPs of the Social Democratic Party and trade unions – were not influential. In economics, the mainstream supported this new pension strategy.

Trends and policy debates from 2000

In Britain, the dominant concern of policy makers since 2000 has differed sharply from that of pensioners, organised labour and most academic experts and think tanks. Reluctant to prevent poverty by increasing the basic pension, policy makers have used means-tested benefits instead. An ideological preference for private pensions has continued despite growing evidence of market risk and fraud. Critics have emphasised rising pensioner inequality and poverty and the disadvantages of means testing.

In Germany, limiting the contribution rate of the social pension scheme (to 20 per cent until 2020 and 22 per cent until 2030) became the dominant political objective, while the pension level became the dependent variable. This shift from a defined benefit to a defined contribution approach occurred in both public pensions and private occupational pensions). The new political strategy has brought:

- a reduction of the burden on public budgets;
- a reduction of labour costs of employers;

- but an increase in the total cost to private households of payments to public and private schemes (the well-known double taxation due to a shift from PAYG to capital funding);
- the introduction of fully taxable pensions whether from social insurance, private life insurance or occupational schemes.

It can be expected that the social pension insurance will be transformed in the long run in fact from a scheme with a relatively close contribution–benefit link to a highly redistributive scheme.

Reforms to private pensions

Acknowledging the limited access of workers to occupational pensions and the exorbitant charges in personal pension schemes, the British government introduced Stakeholder Pensions (SHPs) in 2001. These are a more regulated form of personal pension, provided by the same companies but intended for individuals with modest to average incomes. Employers without an occupational pension scheme must offer an SHP to their employees. Administration charges were capped at 1 per cent per annum of the fund but this was raised to 1.5 per cent following pressure from pension providers. Hidden fees (e.g. for dealing) are not capped, so costs remain high. Takeup of SHPs has been unimpressive, consisting mainly of transfers from less regulated personal pensions. This reflects mistrust of personal pensions following mis-selling and the fall of investment returns after 2000. In fact, employees have been switching back into the State Second Pension (S2P).

Germany also introduced reforms to subsidised personal pensions and to the occupational pension system. Personal pensions will be subsidised if they fulfil certain conditions like a guarantee that savings invested will at least retain their nominal value after costs and that accumulated assets at retirement or on becoming disabled will in principle be annuitised. The subsidy is introduced step by step. From 2008 on, up to 4 per cent of individual earnings can be subsidised. The tax (or transfer) incentives are targeted primarily on people with low earnings and with children. It is however questionable whether many low-income households can afford to save, while the financing of the subsidies will also burden these households (by indirect taxation). High-income households may shift savings to this subsidised type of pension without increasing their individual savings for old age.

Regarding occupational pensions, employees are legally entitled to pay up to 4 per cent of their earnings (up to the ceiling for social insurance contributions) into a DC company-based pension (earnings conversion). This pension is, however, more like a personal than a traditional occupational pension (financed in Germany mainly by employers and being DB). In addition a fundamental change in taxation of pension provision will be phased in. Social insurance pensions, which are only partly taxable now, will become fully taxable, while employers' contributions to social pension insurance will reduce taxable earnings.

Reforms to state pensions

The major British reform was the replacement of SERPS in 2002 by S2P. This pension is earnings-linked but redistributive, intended for those with below average earnings. Limited credits are allowed for family caring. However, the redistributional effects of S2P, as entitlements grow in the future, will be offset by the decline in the basic pension, if it remains indexed to prices.

Government projections indicate that state pension transfers (basic and SERPS/S2P) will fall from £34bn to £26bn (in 1997 earnings terms) between 2000 and 2050 (or from 4.4 to 3.4 per cent of GDP). The combined employer and employee NI contribution rate is expected to fall to less than 15 per cent by 2050. Yet over the same period, the population aged over 65 is expected to rise from 19 per cent to 30 per cent (see Table 2.2) so that by 2040, each pensioner's share of GDP will be 40 per cent less on average than for current pensioners (Pensions Policy Institute, 2003a). The basic pension was projected in 1998 to decline to only 7 per cent by 2050 and the value of the combined state pensions (basic and S2P) was projected to fall from 37 to 20 per cent of average male earnings (Pensions Provision Group, 1998). These figures

Table 2.2 Projected changes in age structure, employment and state pension spending, 2000–2050

Measure	Britain		Germany	
	2000	2050	2000	2050
% aged 65+ among 15+ population	19	30	19	33
% not employed among 15+ population	41	49	45	51
State pension spending* (age 55+ as % GDP)	4.4	3.4	12	17
Spending on means-tested benefits for retired (% GDP)	1.0	2.6	n/a	n/a

* Excluding means-tested benefits in Britain; including civil service pensions in Germany
Sources: Calculated from data in Economic Policy Committee (2003); Government Actuary Department (2003).

were modified by above-inflation rises in the basic pension in 2001 and 2002 following protests at small increases. The trend up to 2007 was to replace state pensions with means testing.

In Germany, reforms to the public pension scheme were decided upon in 2001 and 2004 by the new coalition government to achieve the target level of contributions. The reforms included reduction of the pension level by reformulating the pension adjustment formula twice: increasing the contribution and tax burden of pensioners, and increasing the pensionable age after unemployment. A special form of social assistance for the elderly, a means-tested 'needs-based pension', was introduced.

The range of German pension schemes was extended by this means-tested special transfer payment for the elderly and the subsidised personal pensions. Beside these changes, the 'earnings points' accumulated during unemployment were reduced and there were changes in long-term care and health insurance. In addition, some reductions affected pensioners. However, a comprehensive approach regarding all these effects does not exist in the German public debate on old-age security. Figures 2.1 and 2.2 show respectively the outline structure of the British and German pension arrangements.

Pensioner poverty

Occupational pensions, previously deemed a 'success story' in Britain, have come under pressure from rising longevity and poor returns. Some employers boosted their contributions to compensate but others have limited their risk by switching to DC plans, while reducing their contributions so that projected pensions are some 30 per cent lower. Many DB schemes have been closed to further accruals, while others have been wound up, leaving workers with little pension after years of contributions. Between 2000 and 2004, about half the members of private sector DB schemes experienced closure of their scheme to new members, while about a third saw other reductions in their scheme (Government Actuary Department, 2005). A low-budget Financial Assistance Scheme was set up to provide partial compensation for lost pensions and from April 2005 some compensation will be provided from a new Pension Protection Fund. Like the Pension Benefit Guaranty Corporation in the US or the German Pensionssicherungsverein, this is financed by a levy on firms with occupational pension schemes but it seems likely to be overwhelmed by claims and to require government finance in the future.

Personal pensions and other DC schemes have exposed individual employees to unacceptable levels of risk. This is illustrated by the

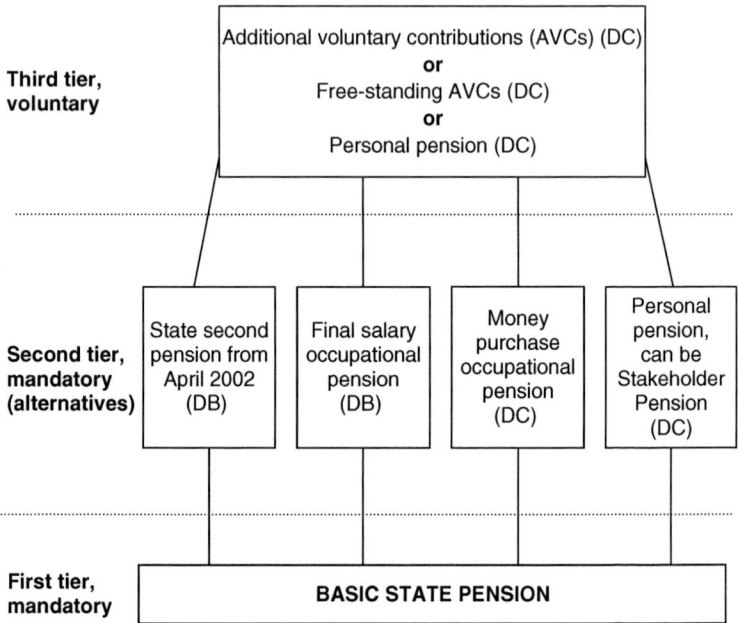

Third tier, voluntary	Additional voluntary contributions (AVCs) (DC) **or** Free-standing AVCs (DC) **or** Personal pension (DC)			
Second tier, mandatory (alternatives)	State second pension from April 2002 (DB)	Final salary occupational pension (DB)	Money purchase occupational pension (DC)	Personal pension, can be Stakeholder Pension (DC)
First tier, mandatory	**BASIC STATE PENSION**			

Notes

DC Defined contribution DB Defined benefit

1. State pensions are PAYG. Personal and occupational pensions are funded (except in the public sector where most are PAYG).
2. Employees earning above the Lower Earnings Threshold (LET, £94/week from April 2005) must pay full NI contributions, but (i) those earning below the Lower Earnings Limit (LEL, £82/week) pay no NI contributions, but may acquire credits through Home Responsibilities Protection; (ii) those earning between LEL and LET acquire NI credits. The self-employed contribute only to the basic pension.
3. NI pays for other benefits such as maternity, sickness, unemployment, and some disability benefits, as well as contributing to the cost of the National Health Service.

Figure 2.1 Outline structure of British pension regime, 2005

collapse of Equitable Life, in which many lost much of their life savings. Moreover, the cost of providing for retirement through private pensions is high. For a man, a private DC pension replacing two thirds of final wages requires paying 24 per cent of earnings continuously from age 25 to 65; for women even higher contributions are required (Mercer, 2002).

Despite the fall in share prices, distrust of personal pensions by workers, and employers' retreat from occupational pensions, the Labour government continues to focus on increased saving through private pensions, rather than through state pensions or other forms of saving.

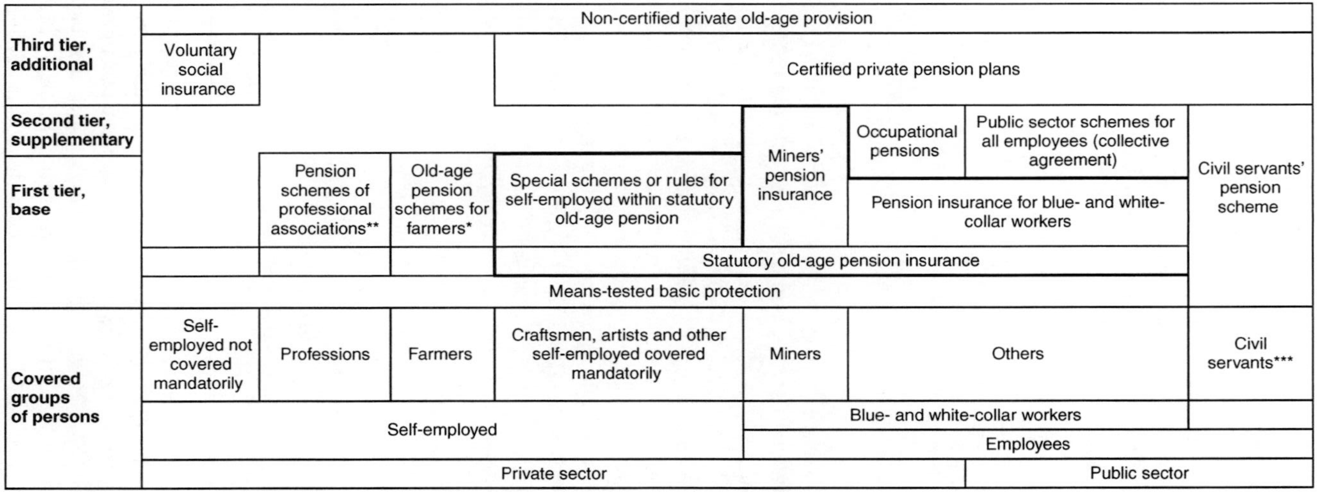

	Non-certified private old-age provision			Certified private pension plans			
Third tier, additional	Voluntary social insurance						
Second tier, supplementary				Occupational pensions	Public sector schemes for all employees (collective agreement)		Civil servants' pension scheme
First tier, base	Pension schemes of professional associations**	Old-age pension schemes for farmers*	Special schemes or rules for self-employed within statutory old-age pension	Miners' pension insurance	Pension insurance for blue- and white-collar workers		
			Statutory old-age pension insurance				
			Means-tested basic protection				
Covered groups of persons	Self-employed not covered mandatorily	Professions	Farmers	Craftsmen, artists and other self-employed covered mandatorily	Miners	Others	Civil servants***
						Blue- and white-collar workers	
		Self-employed				Employees	
		Private sector					Public sector

* Including family workers. This scheme is designed as partial old-age security beside income from the former farm.

** Partly also for employees of the respective branches.

*** Including judges and professional soldiers.

Source: Schmähl (2004).

Figure 2.2 Outline structure of German pension regime, 2005

In Germany, the pension reforms since 2000 were intended to increase saving through occupational and other private pensions. As the government stated, 4 per cent of the gross earnings of individuals is required to compensate for the reduction of the public pension. It seems peculiar that in Germany occupational and other privately funded pensions were strongly promoted during a time when the risks of such provisions (e.g. market failure and fraud) became obvious worldwide. In the discussion about the legal reform, its advocates kept silent about the risks of the capital market. Even in 2005, the development of the capital market was not considered in the discussion on reform of the public old-age pension system. Nor is the experience of other countries, such as Britain, taken into account. Only the positive aspects of privatisation are presented while the consequences for pensioners are ignored. The few dissenting voices attract no attention in the academic community, the media, or among politicians. It is noteworthy that as a result of the pension reforms, *ceteris paribus* the employee contribution rate in 2030 will be 3 percentage points higher and for employers 1 percentage point lower.

Whether the pension reforms will increase private saving in Germany is uncertain. Even with the high subsidy, additional saving of up to 4 per cent of gross income is unaffordable for households with low income. Therefore many of those who wish to save will be unable to do so. Furthermore, people will change their savings portfolio to take advantage of subsidies or of deductions from taxable income. Such substitution effects will not lead to a higher total savings rate. Even if people save more, at first this only increases financial capital to fuel the stock market. But there is no certainty that this will increase real capital as well.

Pension controversies

A number of pension controversies were being debated in Britain and Germany as of 2005.

How should pension policy adapt to longer lives?

The Pensions Commission, appointed in 2002 to advise the British government on pension policy, warned that living longer means paying more into pensions, working longer, or accepting lower pensions. The government favoured longer working, at least up to state pension age, and greater saving through private pensions. Others argued for paying more into state pensions instead since these are fully portable, cost little to run, provide a defined benefit and can redistribute towards those with caring responsibilities and the low paid. Increasing state pension age further was generally rejected as unfair to the poorest, whose life expectancy is lowest.

In Germany, it was decided by the government to adapt to longer lives by accepting lower pensions, as the new pension adjustment formula gradually reduces the replacement rate. As a result, both future and current pensioners are affected. Whereas the first group may be able to compensate for the reduction – especially if they are young – this is impossible for those who already receive a pension. The extent of the reduction can be judged by considering a pensioner who has paid contributions on average earnings over 45 years. Previously the individual could expect a pension of 70 per cent of average net earnings but under the new formula, there is no target replacement rate. Nevertheless calculations indicate that in 2030 the replacement rate will be only 52 per cent. By then, most people will not have the full 45 years of employment. In addition to women with career gaps for caring, many men will have interrupted working careers due to high rates of unemployment. Therefore for the majority the actual pension will be even lower than 52 per cent of actual average net earnings. It is likely that the proportion of older people needing social assistance will increase considerably.

In November 2005 a new 'grand coalition' of Christian and Social Democrats decided to increase the age for a pension without reductions from 65 to 67 up to the year 2030. That means that at age 65 the pension of an average earner after 45 years of employment will only be 48 per cent of average net earnings. The official poverty line in Germany is about 40 per cent of average net earnings. Even an average earner, then, needs nearly 38 years of contribution payment just to get a social insurance pension not as high as means-tested social assistance.

Should tax relief and rebates be ended?

In both Britain and Germany, public subsidies to private pensions, in the form of tax relief on contributions and (in Britain) rebates for contracting out of part of the state pension scheme, are not reported as pension spending – yet are a significant burden on the public budget. In Britain, tax spending on pensions grew from £1.2bn in 1979 to £8.2bn in 1991, reflecting the spread of personal pensions, and by 2000 had risen to £13.7bn, equivalent to over 40 per cent of state spending on the basic NI pension (Sinfield, 2002). Tax relief and rebates are estimated to cost Britain 2.4 per cent of GDP in net forgone revenue (Pensions Policy Institute, 2003b). Such spending is highly regressive, with half the benefit received by the top 10 per cent of taxpayers and a quarter by the top 2.5 per cent (Agulnik and Legrand, 1998). New legislation has, under the pretext of simplification, raised the limits for tax-relieved contributions.

Increasingly, the logic of tax relief on private pension contributions is being questioned in Britain. First, tax spending is targeted on the well-off, who need no incentive to save, while there is no tax relief on contributions to S2P, to which the lower paid contribute. Second, the resources spent on tax relief could greatly improve state pensions. It is hard to justify such a large and regressive public subsidy to private pensions, when no convincing case has been made for favouring private over state pensions.

In Germany, tax relief and subsidies are seen as very important for the transformation of the pension system. The intention is to increase the overall saving rate, seen as impossible without incentives. Private pension providers use these incentives to promote their products to new clients ('government is presenting you with a gift'), meanwhile attacking the public pension system as unsustainable. While it may be more important for future economic development to give more incentives for increasing human instead of financial capital, this is not a topic of discussion in Germany.

Should employers be compelled to contribute more to pensions?

Reductions in employer contributions to occupational pensions led the Trades Union Congress (TUC) in Britain to demand that employers be compelled to contribute more to either occupational or Stakeholder pensions. But the case is equally strong to compel employers to make higher NI contributions. In Germany, trade unions tried to negotiate collective agreements on participation of employers in financing pension claims based on earnings conversion (contributions) of employees. Maybe if there is not enough voluntary saving for occupational and personal pensions, the demand for mandatory occupational pensions will gain more support.

However, in both Britain and Germany, increased compulsion on employers would go against the economic policy goal of promoting growth and high employment through reducing labour costs.

Evaluation of recent pension policy

Pension policy can be assessed on four criteria: incomes adequacy for pensioners; degree of inequality of pensioner incomes; sustainability of pension schemes, and maintenance of incentives to save.

Adequacy of incomes

The EU's first objective under the Open Method of Coordination, set up in 2000, concerns adequacy. In the 2001 Laeken accord, Member States agreed to 'Ensure that older people are not placed at risk of poverty and can enjoy a decent standard of living; that they share in the economic well-being of their country and can accordingly participate actively in public, social and cultural life'.

Britain has failed to meet this objective, given the high proportion of older people in poverty – one fifth – and the growing proportion of pensioners dependent on means-tested supplements – about half. Older women are particularly likely to have inadequate incomes; of the 1.9 million pensioners forced to rely on means-tested benefits in 2005, 1.3 million were women.

The German public pension system has been more successful in the past in providing adequate pensions for many pensioners, due to earnings-indexed social insurance pensions. There remain differences between men and women. Just over 10 per cent of pensioners are poor but this will change in the future. Because of the current pension formula, pensions will only increase if gross earnings increase by more than 1.3 per cent per year. It can be expected that pensions will remain constant in nominal terms for a number of years. In addition the tax and contributions burden of pensioners was and will be increased, reducing in some years even the absolute nominal amount of pension benefit. Only about 50 per cent of employees contribute to an occupational pension scheme.

Inequality of incomes

Income inequality among pensioners is structured by gender, class, marital status and age group, as well as ethnicity, with white middle-class men having the highest incomes (Ginn, 2003; Arber and Ginn, 2004; and see Table 2.3).

Gender inequality in average income is similar in the two countries, with older women's income 57–60 per cent of men's. Married women have the lowest personal incomes (Table 2.3). Divorced women – a growing group – are in a precarious financial position: their income is higher than that of married women but they have no prospect of inheriting a widow's pension. In both Britain and Germany, substantial class differences in income are evident, with the most privileged class grouping having over twice the income of the least. Among men, average income declines with age, due partly to lower private pension income in earlier cohorts, but among women this is not so, as advancing age increases the proportion of widows with a survivor's pension.

Table 2.3 Gross individual income per week by marital status, occupational class and age group, men and women aged 65+ (€, median for Britain, mean for Germany)

Factor	Britain			Germany		
	Men	Women	W/M%	Men	Women	W/M%
Marital status:						
Married/cohabiting	246	81	43	414	143	34
Single	187	157	85	338	307	91
Widowed	207	161	78	932	307	78
Divorced/separated	180	132	74	405	246	61
Occupational class*: Britain						
Professional/managerial	287	148	52			
Intermediate non-manual	142	99	70			
Routine and manual	136	89	65			
Occupational class*: Germany						
Workers				319	212	66
Salaried employees				458	285	62
Civil servants				674	655	97
Self-employed (not farmers)				388	364	94
Co-operators				252	183	73
Age group						
65–69 (Germany 65–75)	255	130	51	356	210	59
70–74	242	132	55			
75–79 (Germany 75–85)	213	132	62	374	243	65
80–84	206	134	65			
85+	177	132	75	323	245	76
All	232	132	57	408	245	60
N	1,474	1,882		13,807	15,142	

* Based on own previous occupation
Sources: Britain: *General Household Survey, 2001*; Germany: Bundesministerium für Arbeit und Sozialordnung, 2001s (authors' calculations).

The distribution of incomes among British pensioners is profoundly influenced by private pensions. Whereas state pensions are redistributive, especially in making some compensation for unpaid caring work that has restricted employment or in case of unemployment, private pension income reflects differential access to a good occupational scheme, as well as earnings and years of employment (Ginn and Arber, 1993). Women, with their greater reliance on state pensions, are the main losers from the privatisation of pensions. In Germany, too, occupational and personal pensions are an important source of inequality of income in old age.

At the turn of the century, private pension income was received by only 43 per cent of older British women compared with over 70 per cent of men (see Table 2.4). In Germany, too, older men were three times more likely than women to have any private pension, 32 per cent compared with 10 per cent. Class differentiation in private pensions is also evident, as in Britain. Because of its dominance in Germany, the public earnings-related pension scheme is the main source of pension inequality, although redistributive elements modify the link to previous earnings. In Germany pensioners with higher public pensions also usually receive higher occupational pensions.

The paradigm shift in the German pension system will lead to growing inequality of pensioner incomes in future. As private pensions become increasingly important, the distribution of pensions will change not only because of more private pension contracts but also because many people will have no private pension entitlement or only a small amount. Under current British pension policy, inequality in pension incomes will increase as state pensions decline and only a privileged minority receive good occupational pensions.

Sustainability of pension schemes

Alarmist accounts have predicted that population ageing will make PAYG pensions unsustainable. However, projected population ageing is relatively modest in Britain over the next decades, the proportion of older people rising by about 1 per cent every five years on average (see Table 2.2). This increase is comparable to that experienced in the past. However, the ratio of pensioners to contributors is a better indicator of pension scheme sustainability than age structure. Rising employment rates among women and older workers help to sustain PAYG schemes. In Britain, employment of these groups has been increasing since the 1990s and is higher than in Germany. For example, British women's employment rate rose from 55 per cent in 1983 to over 70 per cent in 2005, whereas in Germany women's employment rate rose from 51 per cent to 65 per cent in 2004. The employment rate of older German workers is low compared to other countries, at about 59 per cent (age 55–59) and only 22 per cent (60–64) in 2002.

Britain's NI Fund has had a growing surplus of contribution income over benefit spending, due to 25 years of wage-linked contributions but price-linked benefits. The surplus was £38.5bn (€57bn) in 2007, or about half annual spending on state pensions. This surplus, which grows by about £4bn (€5.9bn) every year, is borrowed by the government at low interest rates, for general spending. Therefore, far from being unsustainable, the state NI schemes actually subsidise government

Table 2.4 Private† pensions by marital status and occupational class, men and women aged 65+

Factor	Britain				Germany			
	(a) % receive		(b) median amount for those with some		(a) % receive		(b) mean amount for those with some	
	M%	W%	M €/wk	W €/wk	M%	W%	M €/wk	W €/wk
Marital status								
Married/cohabiting	74	28	132	49	33	7	104	47
Single	52	61	94	101	47	34	52	78
Widowed	70	56	88	66	27	12	74	54
Divorced/separated	57	36	112	69	39	27	70	65
Occupational class*: Britain								
Professional/managerial	90	64	248	137				
Intermediate	60	51	121	62				
Routine and manual	62	34	72	40				
Occupational class*: Germany								
Workers					34	10	45	33
Salaried employees					49	16	138	69
Civil servants					1	5	77	75
Self-employed (not farmers)					5	3	128	24
Co-operators					11	2	7	35
All	71	43	83	44	32	10	97	57
N	1,474	1,882	891	694	13,807	15,142		

† Occupational or personal pension, including survivor pensions

* Based on own previous occupation

Note: The German data concerning the percentage receiving are based on cases. It is possible to receive a pension from private industry and from the public service at the same time so the percentages are biased and only give the upper limit.

Sources: Britain: *General Household Survey, 2001*; Germany: Bundesministerium für Arbeit und Sozialordnung, 2001 (authors' calculations).

spending. Despite this, demographic change has been used to legitimate pension privatisation.

In Britain, as in the US, private pension schemes have too often proved unable to meet their liabilities, due to ageing of scheme membership, contribution holidays by employers during boom years and poor stock market returns. There is thus no reason to suppose that privately funded pensions are more sustainable than state PAYG pension schemes.

It is political sustainability that is a problem in Britain – given the pensioner poverty rate, escalating means testing, disincentives to save and the retreat from private pensions by both employers and employees.

In Germany, 'financial sustainability' of the pension system was mainly used as a vogue expression. It was utilised in the same manner as 'international competitiveness', 'intergenerational equity', 'globalisation' and 'labour costs' to give arguments for the restructuring not only of the old-age pension system but also of the whole social security system. Financial sustainability was used to justify abandoning the paradigm of a stable living standard in old age and introducing the new paradigm of a maximum contribution rate (22 per cent in 2030).

Maintaining savings incentives

Means testing older people creates a pensions poverty trap in which pension income or savings may bring no financial gain because of withdrawal of benefits. This penalises past thrift, discourages earning in retirement and is a disincentive to saving during working life.

In Britain, the basic state pension was about £32 per week less than the minimum guaranteed income in 2007, so that many with a small second-tier pension gained no advantage from it. The tapered Pension Credit reduced the rate of withdrawal of benefits from 100 to 40 per cent, although for women lacking the full basic pension and for tenants the withdrawal rate may still reach 100 per cent. The Pension Credit extended means testing to half of pensioners and is expected to include three quarters by 2028. 'The Pension Credit does not resolve the incentive problems of the MIG [Minimum Income Guarantee]: instead it spreads them out over a much larger group' (Brookes *et al.*, 2002, p. ii).

For many employees retiring around 2050, the reduced basic pension (worth only 7 per cent of average earnings by then) plus S2P will be insufficient to avoid the need for means testing, on current policy. Thus policy makers have yet to solve the problem of ensuring that saving (through pensions or other financial products) is financially worthwhile. For women, decisions about saving are even more complex than for men, due to their typically fluctuating earnings over the life course.

In Germany, too, the pensions poverty trap will become an increasing concern, due to the decline of the replacement rate. A contribution of 20 to 22 per cent will not guarantee a pension above the social assistance level and this problem is not confined to those with very low-earning careers. For example, one has to pay contributions for about 35 to 40 years on average earnings to obtain a pension exceeding the level of social assistance (Schmähl, 2004, pp. 174ff.). And if people realise what they have to save to reach an 'adequate' pension income through private pension schemes this situation will become even more serious, especially for the low paid. In addition, during long spells of unemployment, accumulated assets may be used to finance current spending, reducing assets for the time after retirement. The pensions poverty trap has been growing in Germany since 2000 and it will worsen in future.

Challenges for the next ten years

The main focus of fiscal policy in both countries is to limit public pension expenditures, independent of which political party/parties is/are in power. To increase the tax base, maintaining full employment, including that of older workers, is required. Both policy makers and special interest groups, which have massively promoted the dismantling of the state pensions, would like to see expansion of privately funded pensions. A public debate is needed in which the crude portrayal of 'private good, public bad' – which is used in relation to all kinds of welfare provision – is reconsidered.

In Germany the debate on adequate pension provision does not take into account that important changes also took place in sickness and long-term care insurance, affecting income in old age. While the value of income after retirement may decrease (for example because private pensions are not adequately indexed), costs in case of illness, etc. may increase with age because of growing age-specific mortality.

Increasing saving through private pensions presents formidable problems, as the British government has discovered. Neither occupational nor personal pensions are in a good position to fill the gap created by declining state pensions. Employees on modest incomes generally cannot save more than at present and it is doubtful whether those who could save more should do so through risky personal pensions. Occupational pensions are in decline, except in the public sector, and are unavailable to many workers.

In Britain, the main challenges are to ensure that state pensions lift the majority of pensioners out of means testing, reducing the poverty rate and diminishing income inequality in later life. For those of working age, confidence in adequate state pensions is required to make additional voluntary saving worthwhile. The pension position of women must be urgently addressed by inclusiveness measures to improve their basic pension entitlements. Some have suggested a residence-based Citizen's Pension to achieve this, while others favour retaining the contributory NI schemes. To meet these challenges, more resources must be put into pensions, and employment rates must be increased, especially by expanding quality, affordable child care services and creating suitable job opportunities for older workers. A lasting settlement is needed, to restore confidence in the reliability of state pensions.

In Germany, the major challenges are the same overall. For the next ten and more years reducing unemployment is highly important. Public pensions depend mainly on working, and saving – whether for public or private pensions – is only possible if people can afford to save and build entitlements. Therefore a further challenge is to guarantee wages high enough to enable saving for retirement. In general, it should be possible for people to maintain their standard of living during retirement. Pension income should be substantially above the official poverty line.

Conclusion

Britain and Germany come from contrasting welfare traditions, the latter relying more on public pensions, yet in both countries state PAYG pensions are under attack as policy makers promote further privatisation. In Britain, this emphasis prevails despite workers' loss of confidence in private pensions following evidence of market failure, fraud and incompetence. The effects of increasing private pension provision at the expense of state pensions are evident: a relatively high rate of pensioner poverty, especially among women, very unequal incomes and a majority of pensioners subject to the indignity of means testing. Working-age people cannot plan sensibly for retirement, faced with the pensions poverty trap. Moreover, the cost of private pensions in forgone tax revenue has escalated, targeting benefits towards the highest earners and reducing public resources needed for state pensions. Ignoring this British experience, a consensus in Germany among the mass media, politicians and many economists has been in favour of a shift to more capital-funded private pensions. Policy makers and some economists have demanded a cut in public pensions to improve intergenerational equity and sustainability (Schmähl, 2005a).

The claim that population ageing requires a shift from state to privately funded pensions has been questioned by pension analysts. Some writers have examined the issue of intergenerational equity from an ethical standpoint, concluding that fair sharing between generations is better achieved by state PAYG pensions (Schokkaert and van Parijs, 2003). Many economists acknowledge that private pensions are adversely affected by population ageing, leading to funding deficits, defaulting on pension promises and financial risk to companies. Thus it is not clear why private pensions should be preferred to state pensions. Some social policy analysts argue that a crisis in state pensions has been created in order to present a political choice as an economic imperative (Walker, 1990; Vincent, 1999). They suggest privatisation is motivated by ideological opposition to public welfare: 'Political ideology has distorted and amplified the macroeconomic consequences of population ageing in order to legitimate anti-welfare policies' (Walker, 1990, p. 377).

For EU countries beginning to promote private pensions, it is possible – but unlikely – that the consequences of privatisation seen in Britain may be avoided. The danger is that, as private pensions grow among the middle classes, this will create the political conditions required for making more severe cuts in public pensions, resulting in increased pensioner poverty and inequality.

Appendix: Pensioner poverty

The poverty rate among pensioners is twice as high in Britain as in Germany. The proportion of Britons over 65 with a household equivalised income below 60 per cent of the population median rose between 1981 and1991 from 16 to 29 per cent, fell back to about 20 per cent by 1996 and since then has remained at around this level (Office for National Statistics, 2002).

The British basic state pension for a single pensioner in 2005 was £82 (€118) per week if received in full. Most men receive the full amount but only 30 per cent of women do so. Among women pensioners, nearly two thirds receive less than £62 (€89) per week (Pensions Commission, 2004, Figure 8.10) because most older married women rely on receiving 60 per cent of their husband's basic pension. Younger cohorts of women will increasingly be entitled to the pension on their own contribution record. The amounts of SERPS or S2P received by pensioners are relatively small, especially for older women. Because of the low amounts of British state pensions, a combination of full basic and full S2P combined (at maturity in 2050) will for most women provide a pension below the level at which means-tested benefits are payable (Falkingham and Rake, 2001).

The British government's response to pensioner poverty has been an unprecedented expansion in means testing. Yet this has failed to reach many of the poorest, who fail to claim due to the stigma, intrusiveness and complexity of the process. Takeup remains between two thirds and three quarters of those eligible, despite government campaigns. The cost of means-tested benefits for pensioners is projected to rise steeply while pension spending falls, on current policy (Table 2.2).

In Germany, pensioner poverty is lower than for other age groups, only 11.4 per cent (Bundesregierung, 2004) and is therefore at present not a political issue. This reflects pension policy in the past. Whether this situation will continue in the future is doubtful. With high unemployment since the beginning of the 1980s and unsteady employment since then, many employees will have low entitlements in the public pension system and be unable to save enough money in occupational or other private pensions. This also applies to a large part of the self-employed – whose income distribution is only slightly more dispersed than for employees, limiting their ability to save. Moreover, as it is not mandatory for self-employed workers to have old-age pension insurance, around 70 per cent of them are not covered. In the future, this situation will become even worse as restructuring of the labour force takes place, substituting dependent workers with self-employed. This development is being fostered by economic and fiscal policy.

Increased savings through private pensions

The realisation that both occupational and personal pensions are inherently risky makes increased saving unlikely. In addition to this, if there are subsidised types of pension saving there may be a shift of money from other types (in particular by households with higher income). Among employed British individuals, 67 per cent of men and 53 per cent of women contribute to a private pension (Office for National Statistics, 2004). Including all adults of working age, the proportions contributing are 46 per cent of men and 38 per cent of women.

Acknowledgments

We are grateful to Debora Price for analysis of the British General Household Survey, to the Office for National Statistics for permission to use data, and to the Data Archive, University of Essex, for access to data.

3
Ageing and the Labour Market – A Comparison of Policy Approaches

Frerich Frerichs and Philip Taylor

Introduction

Both Germany and the UK are experiencing substantial ageing of their workforces and, simultaneously, their workforces are shrinking. At the same time it is important to note that older workers, particularly men, have been regarded by employers and policy makers as a reserve labour army in the past in both countries (Naegele and Walker, 2002a). Older workers have been confronted with numerous forms of direct and indirect discrimination in both the workplace and in the labour market in general. The result has been long-term unemployment and non-employment among older workers. Employment rates of older workers in both countries have declined dramatically over the past twenty years, although significant differences between the United Kingdom and Germany can be observed (Walker, 2002a). Low labour market participation rates are mainly due to early retirement schemes in Germany, which have been implemented in past decades (Naschold *et al.*, 1994; Ebbinghaus, 2001) and due to usage of occupational pension schemes, disability benefits as quasi-early retirement, early retirement schemes and discouragement from staying in work in the UK (Taylor and Walker, 1996; Taylor and Unwin, 1999). Although early exit pathways have been terminated or their scope limited and there is an increasing emphasis on prolonging working life, the legacy in terms of promoting negative views of older workers is persistent.

Managing the effects of demographic change on future labour supply, and age-related employment barriers, requires active and integrated strategies. It is now widely accepted that extending working life may not only reduce pension, health and social welfare costs for the state but will also provide the opportunity for a happier and more productive old age for many individuals. However, to achieve this goal, measures have to be taken both at a micro level (within companies) and at a macro level (in the labour market), and these measures require underpinning by public policies.

For some time now, governments in Germany and the UK have been trying to direct more attention to employment opportunities for older workers, beginning to implement policies aimed at increasing labour force participation rates among this group (Naegele, 2002; Walker, 2002a). A new policy consensus seems to be emerging around the notion of 'active ageing'. Both countries are implementing policies aimed at reducing age discrimination in the labour market, encouraging the recruitment of older workers, delaying early exit, encouraging the sharing of best practice among employers, or helping older workers who are seeking guidance or training.

What accounts for this recent shift in policy? There are three main reasons:

- the concept of sustainability of both the welfare state and the public old-age income security system;
- the ageing and shrinking of the workforce;
- labour shortages in some parts of the economy on the one hand, and enduring or even increasing (long-term) unemployment of older workers on the other.

Ultimately, the challenges both countries face against this backdrop of social change lie in the employment prospects of future cohorts of older employees, which will need to be enhanced. At the same time working life will require prolongation. Since the two countries face similar challenges concerning the employment of older workers, the overarching question is if they can and will develop successful measures to integrate older workers. Therefore, this chapter analyses the developments as well as the similarities and differences in policy approaches in the respective countries. In this context the focus is on active labour market policies (ALMP), which refer to direct government intervention in the labour market to prevent or alleviate unemployment. Their main components include job placement, job training and job creation policies. However, to assess the real impact of active measures, passive measures like unemployment benefits and early retirement provisions also have to be taken into account.

Despite similar challenges, labour market policies in Germany and the UK are frequently at opposing ends of the spectrum, both in terms of welfare state typologies and in terms of industrial relations regimes (cf. Hassel, 2001). The German welfare regime has been characterised as 'conservative-corporatist' (Esping-Andersen, 1990), whereas the British case, though deviant in nature, can be labelled as a 'liberal welfare' regime. According to Esping-Andersen (1999, p. 74): 'liberal welfare regimes in their contemporary form reflect a political commitment to minimise the state, to individualise risks, and to promote market solutions'. Concerning labour market policies, this is manifested in needs-based and means-tested unemployment benefits as well as the relatively high weight given to occupational pensions. Furthermore, liberal social policies favour a passive approach to employment management and give priority to unregulated labour markets (Hassel, 2001; Walker, 2002a).

In contrast, conservative labour market policies favour strong job protection for all employed adults and active employment or training policies are more common here than in liberal regimes, but unemployment is also strongly managed by labour supply reduction including, for instance, early retirement. Thus, the conservative-corporatist approach, as exemplified by Germany, was characterised by a practice of very early retirement in the 1980s and 1990s, supported by a strong consensus between enterprise, trade unions, the state and workers. The result was low economic activity rates for workers from age 55 onwards (Delsen and Reday-Mulvey, 1996; Hassel, 2001).

Applying the concept of welfare regimes may help us to consider why certain approaches have been selected at all by governments and why these have been implemented in a certain way. The guiding question in this theoretical context is concerned with how different welfare regimes react to the challenges that demographic changes inflict on both society and the labour market. Is there a great deal of path dependency, due to enduring policy structures and opinions? Or is there convergence of policy approaches that can be traced back to the common, groundbreaking challenge of demography or globalisation?

The basic assumption is that labour market policies for older workers in the UK and Germany are outcomes of both particular welfare state regimes and existing institutional constraints and choices under which the respective governments, unions, employers' associations and individual firms and workers make their decisions. What might look like a similar tendency towards more active labour market policies for older workers in the UK and Germany may be concealing fundamental differences in approach.

Older workers and the labour market in Germany and the UK

Whereas in the past, workforce ageing has been slowed by the early exit of older workers and a lower retirement age for women, this will not be the case in future. The average age of the working-age population is projected to increase from 39.0 years in 2000 to 40.2 years in 2020 in the UK and from 39.6 years in 2000 to 41.5 years in 2020 in Germany, and the proportion of adults aged 50–64 will increase from 27 per cent in 2000 to 32 per cent by 2020 in the UK and from 30 per cent in 2000 to 39 per cent in 2020 in Germany (Rössel *et al.*, 1999; Dixon, 2003). Furthermore, early exit is currently restricted by statutory measures in Germany and will be made increasingly unattractive for older workers in the UK due to changes in rules governing occupational pensions. The lower retirement age for women has already been abolished in Germany and will cease in the UK between 2010 and 2020.

While this might seem to lay common ground for both countries in respect of labour market policies for older workers, it appears that the UK is in a far better position to cope with the challenges ahead because of already higher participation and employment rates and far lower unemployment rates for older workers. Table 3.1 indicates that labour force participation of workers aged 55–64 in the UK is at present approximately 14 percentage points higher than in Germany in the cases of both men and women. There has been a steady and pronounced increase in participation rates in the UK since the year 2000, while in Germany only a slight increase has occurred.

Table 3.1 Labour force participation rates among workers aged 55–64, 1990–2004 (%)

Age	1990		2000		2002		2004	
	UK	G	UK	G	UK	G	UK	G
Both sexes 55–64	53.0	39.8	52.8	42.9	55.2	43.3	58.0	44.2
Men 55–64	68.1	55.9	63.3	52.4	65.0	52.6	68.0	54.8
Women 55–64	38.7	24.7	42.6	33.5	45.7	34.1	48.3	33.8

Source: OECD (2005).

However, ratios were very much lower in the UK in the 1980s and 1990s also. In the meantime, labour market policies or labour market regulation in the UK have not changed so much that this could account for these changes. Rather, the current situation may be attributed to better economic circumstances in general in the UK and also to the lasting influence of the unification process in Germany. It remains true, though, that lower participation rates in Germany were in the past systematically induced by the long-enduring process of early retirement, this being particularly pronounced in a country with a conservative-corporatist welfare regime. In the UK public early retirement policies were abandoned at the end of the 1980s due to financial constraints, and were never widespread anyway (Taylor and Walker, 1996).

However, OECD data on the incidence of part-time employment amongst older workers (OECD, 2004a) indicate that the UK's apparent advantage in terms of employment rates among older workers is reduced when part-time employment is taken into account. Table 3.2 shows that the difference in employment rates for older male workers between Germany and the UK is reduced from 10.9 per cent (unadjusted rates) to 6.6 per cent (adjusted rates), while for women the reduction is even greater, from 13.3 per cent to 4.5 per cent.

About half of the difference in employment rates can be explained by a higher incidence of part-time employment, and so British older workers may not be benefiting from better labour market integration as such. On the contrary, it may indicate a weak attachment to the labour market for some. Furthermore, there is a negative relationship between participation rates of older men and average replacement rates for old age pensions (OECD, 2001, 2004b). Average replacement rates are lowest in the UK when compared to other OECD countries, whereas they are relatively high in Germany – reflecting a typical difference between

Table 3.2 Employment rates of people aged 50–64 before and after adjustment* for hours worked, 2000 (%)

Country	Men		Women	
	Unadjusted	Adjusted	Unadjusted	Adjusted
Germany	57.7	59.0	39.6	29.0
United Kingdom	68.6	65.6	52.9	33.5

*The adjusted employment rate is obtained by multiplying the employment rate by weekly hours worked and dividing by 40.
Source: OECD (2004a).

Table 3.3   Unemployment rates among workers aged 55–64, 1990–2004 (%)

Age	1990		2000		2002		2004	
	UK	G	UK	G	UK	G	UK	G
Both sexes								
55–64	7.2	7.7	4.4	12.3	3.5	10.8	3.1	11.3
Men								
55–64	8.4	7.0	5.5	11.5	4.3	10.3	3.9	10.9
Women								
55–64	5.0	9.1	2.8	13.6	2.3	11.7	2.1	12.0

Source: OECD (2005).

liberal and conservative welfare regimes – and this would also explain higher participation rates in the UK compared to Germany.

Unemployment among older workers is almost four times higher in Germany than in the UK, with respective unemployment ratios in the 55–64 age group in Germany at 11.3 per cent and in the UK at 3.1 per cent in 2004 (Table 3.3). However, over the last decade this has not always been the case. In 1990 – and prior to German unification – the unemployment rates of workers in the 55–64 age group were similar: 7.2 per cent in the UK and 7.7 per cent in Germany. More recently, since 2000, a decrease in the incidence of unemployment among older men and women has occurred in Germany, though rates still far exceed those for the UK.

The complementary side of labour force participation – inactivity – is pronounced in both countries and there are various pathways out of the labour market for older workers (Table 3.4). In the UK, 14.2 per cent of the population aged 50 to state pension age are claiming illness or disability benefits. In Germany the main gateway to inactivity is provided by pre-retirement (28.7 per cent), early retirement programmes continuing to be viewed as a legitimate way of dealing with mass layoffs and persistently high unemployment. However, in the UK also, occupational pensions provide a gateway to early retirement, though to a far lesser extent than in Germany.

Dealing with unemployment and inactivity is the main task of active labour market policies. In this regard it is notable that public expenditure on labour market programmes varies greatly between the two countries. Germany in 2003 spent 3.46 per cent of its gross domestic product (GDP) on labour market programmes, whereas in the United

Table 3.4   Reasons for being inactive, people aged 50–64, 2000 (%)

Country	No work available	Retired	Illness or disability	Family duties	Other	Total
Germany	0.4	28.7	4.1	5.0	7.1	45.4
United Kingdom	0.5	12.8	14.2	5.1	3.9	36.6

Source: OECD (2004a).

Kingdom the respective proportion was 0.93 per cent (OECD, 2005). Although the unemployment rate in Germany is almost double that for the UK, this does not fully account for the marked difference in expenditure. This becomes clearer when referring to the ratio of active spending per person unemployed relative to GDP per person. This ratio for Germany is more than twice as high as it is for the UK (0.16 vs 0.07) (OECD, 2003). Considering programme categories in more detail, differences are even more striking. Expenditure on labour market training in Germany is ten times higher than it is in the UK (OECD, 2005). The same is true for subsidised employment and measures for the disabled. By contrast, the difference in expenditure on public employment services is less pronounced.

Development of policy approaches

Policy approaches in the United Kingdom

The ageing of the labour force is currently high on the agendas of policy makers in the UK. In recent years, numerous official reports have considered the economic implications of population ageing and have emphasised the importance of reactivating older workers and tackling age barriers in the labour market (e.g. Cabinet Office, 2000; House of Lords, 2003). This situation is interesting, given that early retirement as it has been known in Germany and other European nations was never as significant a feature in the United Kingdom.

Now, however, and perhaps even despite not having developed the structure of myriad early retirement pathways of some continental European countries, there is an increasing emphasis on extending working life in policy debates in the UK. Policies are in place aimed at promoting the employment of those with disabilities, while the notion of employment as a source of income among those in the Third Age has also gained credence among policy makers. What accounts for the British government's adoption of a later-exit agenda is unclear but it may in part be due to the factor that a powerful age lobby has campaigned for the right of older people to work. However, the government's Welfare to Work agenda and the so-called 'pensions crisis' may have played the most decisive role.

Furthermore, the current Labour government is less opposed to state intervention in the labour market than the previous Conservative government had been and therefore may play a greater role in reversing the early-exit trend and activating or reactivating older workers (Walker, 2002).

A report commissioned by the government and prepared by the Cabinet Office's Performance and Innovation Unit (Cabinet Office 2000) set out a number of recommendations for the development of policies towards people in the third age. Key recommendations were as follows:

- Government to set out its vision of the role and value of older people in society.
- Increase contact with, and job-search assistance for, people on sickness and disability benefits.
- Provide careers information and advice to older displaced workers.
- Raise the minimum age at which an immediate pension is payable.
- Increase the transparency of occupational pension schemes by showing the cost of early retirement in company accounts.
- Promote the advantages of diversity and flexibility in working practices through a group of 'champion' employers.
- Each Civil Service department to review the case for increasing its retirement age to 65 (previously 60).

Recognising that age discrimination impacts on many areas of policy, the government also created a Ministerial Group on Older People to coordinate work across departments. The Secretary of State for Work and Pensions is the Government Champion for Older People and the Department for Work and Pensions (DWP) the lead Department. DWP has responsibility for monitoring progress by means of updates from other departments and for liaising with them on the development and expansion of a strategic approach to tackling older people's issues. However, a recent report concluded that there is an 'outstanding need to provide an overall framework for work across Government affecting older people because, despite progress in joining up policy-making, there remains a lack of co-ordination in some areas' (Comptroller and Auditor General, 2003, p. 8). The report suggested that this might be assisted by publication of the government's older people strategy, which has recently occurred.

The re-election of the Labour party to government in 2001 brought with it a renewed emphasis on productive welfare policy which places employment at the centre of both social and economic policies (Walker, 2002). The party's election manifesto contained specific commitments in this field: to raise the employment rate from 50 per cent to 70 per cent; to tackle discrimination; to extend New Deal 50 Plus (see below), and to examine ways in which occupational pensions can be combined with part-time work so as to encourage flexibility in retirement. These commitments reflect the explicit European Employment Guidelines on Developing a Policy for Active Ageing (European Council, 2001).

The present government's manifest position on older workers is neatly summarised in its Pensions Green Paper where it set out proposals for reforming the retirement income system. The government sought views on the following issues (DWP, 2002):

- providing additional help for those aged 50 and over and recipients of incapacity benefits to return to work;
- treating men and women between 60 and women's state pension age as active labour market participants when women's state pension age increases from 2010;
- maintaining the state pension age at 65, while providing increases for those deferring the claiming of their pension;
- implementing by December 2006 age legislation covering employment and vocational training;
- allowing people to continue working for their employer while drawing their occupational pension, raising the earliest age from which a pension may be taken from age 50 to age 55 by 2010, and consulting on best practice to ensure that occupational pension rules do not discourage flexible retirement;
- changing public service pension scheme rules, for new members initially, to make an unreduced pension payable from 65 rather than 60.

These proposals were important in that they recognised the need for an extension of working life and the need to support disability benefit claimants in returning to employment.

To oversee much of its work in this area, the government established an Extending Working Lives Division of the Department for Work and

Pensions, which has responsibility for increasing employment rates and the opportunities for people aged 50 and over to save for retirement. The division was established with the goals of encouraging employers to implement policies aimed at the inclusion of older workers, increasing the employment rate of older workers and raising awareness of age discrimination more generally. Furthermore, a role of the division is to develop detailed statements of employer best practice which will provide step-by-step guidance on age management. Another initiative will be the piloting of innovative approaches to encouraging the economically inactive to seek work.

The majority of inactive people aged between 50 and state pension age receive incapacity benefit because of a health problem or disability. It is the government's view that new claimants have expectations of returning to work but fail to make this transition, becoming, in effect, early retired. It is also their view that a lack of skills currency, low levels of confidence and employer discrimination are as important as health difficulties in reducing employment opportunities. In 2002 the government published *Pathways to Work – Helping People into Employment*, a Green Paper on measures to help recipients of incapacity benefits to return to work. Based on this Green Paper, the following pilot projects have been undertaken:

- early active support – ongoing support from a personal advisor in the form of a regime of repeat work-focused interviews, combined with action planning with a strong emphasis on returning to work;
- increasing access to specialist employment programmes – the introduction of better referral arrangements to existing provision such as New Deal for Disabled People and New Deal 50 Plus (see below) and new work-focused rehabilitation support;
- the introduction of a 52-week return-to-work payment of £40 per week to all those moving back from Incapacity Benefit (IB) to work for more than 16 hours per week;
- more support for those having to come off Incapacity Benefit and on to Job Seeker's Allowance (JSA), with mandatory early access to the relevant New Deal to allow this group to get personal support from the outset of their JSA claim.

Overall, it can be concluded that the employment of older workers is now much higher on the agenda of policy makers than it was in the 1980s and the first half of the 1990s. A number of important challenges remain, however. In particular, the government seems to be some way from activating the economically inactive, particularly IB claimants. Given that many such claimants have been so for a considerable time, the likelihood of major breakthroughs in the near future must be doubted.

Policy approaches in Germany

The decline of labour force participation rates among older workers in Germany not only mirrors strategic personnel policy measures of German enterprises, but is also the result of a long-lasting broad 'social consensus' among employers, employees, trade unions and work councils, labour market politicians and local labour market agencies regarding early retirement. Currently, there is a strong tendency in public policy towards cancelling this consensus. Initiatives to stop early retirement, first launched at the beginning of the 1990s, have been reinforced since. With pension reforms in 1992, 1996 and 1999, the German government decided to raise both the general pension age and the age limit for specific types of pensions. From 2005, only the 'standard' age of 65 was valid. Employees who still wish to retire earlier will experience a 3.6 per cent annual reduction in their pension.

As regards developing active labour market polices for an ageing workforce, in 2001 the partners of the so-called 'Alliance for Jobs, Training and Competitiveness', which included representatives from government, trade unions and employers' associations and operated at the federal level, agreed a special programme for promoting the employment prospects of older workers. In so doing, a paradigm shift occurred. For the first time, government and the social partners jointly turned away from a policy of early retirement, focusing instead on explicitly preventing older workers from becoming unemployed and promoting their reintegration. Their joint declaration included the following proposals (Gemeinsame Erklärung, 2001):

- raising awareness among business and workers of the benefits of lifelong learning;
- promoting vocational training for older workers, based on voluntary in-company actions as well as on collective bargaining;
- implementing financial incentives for vocational training of older workers aged 50 and over in small and medium-sized companies;
- lowering the qualifying age for wage subsidies from 55 to 50 years.

Elements of these proposals were incorporated in the Job-Aqtiv-Law which came into force in 2002. The aimed to transform current employment policy, which had mainly concentrated on those who were already unemployed, into one which was preventive. Its basic approach was to promote employment while increasing requirements for the unemployed (i.e. a carrot-and-stick approach). For example, employment offices are now obliged to work out a job profile for every unemployed person based on their skills and work experience. Furthermore, a 'reintegration contract' which records the necessary steps to be undertaken both by the unemployed person and the employment agency to achieve reintegration into the labour market is required. By doing this, the government intends to create the basis for a more 'active' strategy. With respect to older workers, the following measures are of particular importance:

- promotion of vocational training: since 2002, employment agencies are able to fund vocational training for workers aged 50 and over if they are employed in a company with less than 100 employees and if the company pays their salary during participation in training;
- extension of special wage subsidies for older workers in the secondary labour market for West Germany.

In a further advance on this approach, in 2002 a commission called 'Modern Services in the Labour Market' (also called the 'Hartz Commission' after Peter Hartz, human resource manager and member of the board of Volkswagen AG, who was its chairman) was established by the federal government, which proposed a comprehensive reform of labour market policies in Germany. Its main goal was as follows (Hartz *et al.*, 2002, p. 19):

Employment promotion policy will be reshaped into an activating labour market policy with particular emphasis on a personal contribution towards economic integration on the part of the unemployed, a concept which will be both supported and secured by a host of relevant services and support programmes.

The proposals made were multidimensional, comprised of 13 components, ranging from radical reform of public employment services to rather vague notions concerning the participation of the nation's elite. Up to the summer of 2004, these proposals were translated into statutory provisions by the federal government. Altogether, four 'Laws Concerning Modern Services in the Labour Market' have been implemented (referred to as Hartz I, II, III, and IV).

In particular, Hartz I includes a number of elements supporting the position of older workers, which intend to improve their labour market prospects. These include:

- Unemployed persons aged 50 and over or employees threatened by unemployment shall, for a limited period of time, receive a monthly subsidy amounting to 50 per cent of their last net remuneration.
- Employers shall be exempt from contributing to unemployment insurance (3.25 per cent of the gross wage of the employee) if they hire an unemployed person aged 55 and over.
- From 2003 onwards, the age limit for fixed-term employment was lowered from 58 to 52 years, whereby contracts can be terminated without justification and without time limit, in order to improve older workers' prospects of reintegration.

The original intention of the Hartz Reforms was to strengthen labour market services and to facilitate job placement activities. Even more important was a shift in general orientation. Although the Job-Aqtiv-Law stressed that unemployed workers were obliged to take a job, in the wake of the Hartz Reforms a workfare approach became more dominant. The main focus lay on combatting 'welfare dependency' and on recipients' obligation to try to become self-sufficient. The obligation of society towards the 'excluded' (i.e. towards the unemployed who were not able to be placed in employment within a short time frame) became less important. This was reinforced in 2003 when the Chancellor launched a labour market package and social welfare reforms titled 'Agenda 2010', despite rhetoric on the need to balance 'demands and promotion' ('Fordern und Fördern').

The 'passive' side of labour market policy has also been reformed so as to contribute to the overall activating approach. Two major reforms are of particular importance to older unemployed workers. One is the merging of unemployment assistance and social assistance and the creation of the so-called Unemployment Benefit II, also known as 'Grundsicherung für Arbeitslose' (Basic Security for Job Seekers). This new benefit, granted since the beginning of 2005, is a tax-funded, means-tested type of benefit to secure the income of unemployed workers once Unemployment Benefit I is no longer paid or if the qualifying conditions are not met. The duration of this benefit is not limited. However, by contrast with former unemployment assistance, which on average guaranteed an income level of 50 per cent of last net income (53 per cent for married unemployed), payment here is reduced to the level of social assistance. Furthermore, from 2006 onwards, receipt of unemployment benefit for older workers has been limited to 18 months (the previous duration was 32 months).

Overall, the most recent labour market reforms provide an ambiguous picture. On the one hand additional measures have been implemented, such as wage insurance, whereby an approach has been chosen which more actively includes older workers. However, specific problem groups such as long-term unemployed older workers are mostly neglected. On the other hand, employing older workers without providing the necessary resources and without guaranteeing that they will work, while at the same time loosening the social security net, raises suspicions that behind the scenes budgetary reasons prevail. Strong indicators for this can be seen in funding restrictions and participation in creating further training initiatives and job creation measures. A more integrative strategy to avoid these problems, involving such measures as the creation of active job centres, does not seem to be possible in the near future because of a lack of staff resources.

Comparison of policy approaches

Even though the rhetoric in terms of active labour market policies against the background of demographic change is quite similar, and in recent years labour market policies concerning older workers in Germany and the UK have converged somewhat as a result of the election of Labour governments in both states, the systems behind these reforms remain quite different. As Knuth *et al.* (2004) point out, active labour market policy in the UK in general started from a very low level of intervention and social security, and had yet to reach the most disadvantaged inactive older people. For example, policies in place aimed at promoting the employment of those with disabilities were in their infancy and still at the pilot stage. Analysis of public expenditure on labour market programmes in Germany and the UK has shown that the ratio of active spending per person unemployed relative to GDP per person is twice as high in the former as it is in the latter.

These differences can be traced back to differences in the welfare state regimes of these countries. The UK has no recent tradition of active labour market policy. Furthermore, policy reversal in the UK as a liberal-residual welfare state is largely driven by long-term concerns about demographic trends and about welfare state disincentives to work, since early retirement was less a pressing structural problem and public pensions have played a far less significant role. Nevertheless, the current Labour government is less opposed to state intervention than previous Conservative governments, emphasising a productive welfare policy with respect to older workers and increasing employment rates in this age group.

By contrast, Germany starts from a quite different position. Concern about sustainability of the pension system is widespread and thus attempts to limit early retirement and reactivate older workers are current policies. In the wake of the recent labour market reforms – Job-Aqtiv-Law and the so-called Hartz Reforms – additional measures to integrate older workers into the labour market such as wage insurance, special training grants, and a shift in paradigm towards a more activating approach to job placement have been implemented. This builds upon a wide array of active labour market measures regulated at the federal level by the Social Security Code which have been in place for more than 35 years. However, it has to be stated that more recently, there is a gradual shift from 'active' to 'activating' policies which goes hand in hand with reducing budgets for active measures.

Where on the one hand, Germany seems to have a broader array of active policies, on the other hand the UK apparently has a somewhat more coordinated or 'joined-up' approach with such organisational features as the establishment of the Extending Working Lives Division at the Department for Work and Pensions and of a Ministerial Group on Older People to coordinate work across government departments.

In this context, it is intended to develop and articulate an evidence-based strategy for extending working lives covering central issues related to employment policies for older workers. In Germany, no such structures exist, with the initiatives and legal measures of the different ministries involved, such as the Ministry of Economy and Labour, the Ministry of Education and Research, and the Ministry for Health and Social Affairs not being centrally coordinated and repeating or even contradicting each other's initiatives. In general, there is no explicit and integrated public 'older workers policy' in Germany so far. This is despite the fact that the 'Alliance for Jobs, Training and Competitiveness' launched a skills strategy for older workers which partly became enacted in the Job-Aqtiv-Law and despite older workers being one main target group in Social Security Code III, the former Employment Promotion Law. An ad hoc, fragmented policy approach still prevails.

Active labour market measures for older workers

Active labour market measures in the UK

New Deal 50 Plus

Active labour market measures in the UK frequently feature older workers as a target group. New Deal 50 Plus, New Deal 25 Plus and Work Based Learning for Adults (WBLA) are measures of significance. Job creation, promotion of self-employment and other measures like work trials are negligible in terms of numbers of participants or virtually non-existent. The most important of these measures is New Deal 50 Plus, which has been utilised by large numbers of older workers, although those on 'inactive' benefits have been much less likely to participate.

Furthermore, in the UK, job centres are responsible for supervising – and, to a lesser extent, delivering means to assist – the active job search required by the Job Seeker's Allowance (unemployment benefit) regime, which is rather strict. The New Deals described below are supplementary to this and target help on those who are hardest to reach. Job centres and New Deals, in contrast to German employment agencies, have three features that offer, in theory, better job placement support to older workers:

- 'Job entry targets' provide extra credit points for job agents if they place the long-term unemployed.
- There is a ratio of 1 job agent to 40 clients in New Deal programmes.
- Vacancy service managers and key account managers take care of vacancies offered by employers.

New Deal 50 Plus was implemented in 2000, aiming to provide practical assistance and support to help people aged 50 or over compete more effectively in the labour market. It can be seen as an integrative measure because it combines support for job seeking with in-work income support, support for training and work trials. In its original form (Department for Education and Employment, 2000) it offered employment advice to non-employed older people who had been claiming benefits for at least six months and who wished to return to work. The programme was voluntary (unlike other elements of the New Deal programme) and open to people inactive on benefits as well as the registered unemployed. A wide range of practical help from a personal advisor was available: help with job search skills, costs of travel to interviews, work-based learning for adults and work trials. There was also a range of help specifically for people with disabilities. It paid an Employment

Credit – an extra £60 a week, tax-free, on top of a person's wage if they took a full-time job (£40 for a part-time job). This top-up money was paid directly to the employee – not the employer. New Deal 50 Plus guaranteed a take-home wage of at least £180 a week (over £9,300 a year) for the first year of work in the case of full-time employment. The Employment Credit could also be used to help set up a small business. An in-work training grant of up to £750 was also available. The programme has subsequently been revised and now has the following main features:

- personal advice and support in finding a job;
- a £1,500 in-work training grant;
- access to financial support when in work, paid as a Working Tax Credit, the amount depending on individual income and circumstances.

The Working Tax Credit was introduced in April 2003 and includes a return-to-work element for people aged 50 and over who have been receiving some out-of-work benefits for at least six months. The aim here is to provide additional help for those facing barriers to returning to work. However, there are concerns that the introduction of the Working Tax Credit might have negatively affected takeup of New Deal 50 Plus. The Employment Credit had been a much more tangible incentive, while the Working Tax Credit must be applied for, and there may be a delay in receiving it; household circumstances will be taken into account in the application, and the employer will be aware of the claimant's circumstances.

Takeup of the programme was lower than had been hoped. Since its launch in April 2000, over 110,000 clients have moved into employment via the programme (Department for Work and Pensions, 2004), although this represents a minority of those classified as unemployed and only a small fraction of those who are actually eligible. Of these starts, a third had a disability, and almost a third were women. Reports from the Employment Service show that there was a steady increase in takeup of New Deal 50 Plus Employment Credit from April 2000, although recent data demonstrate that participation has gradually tailed off since around the time of the introduction of the Working Tax Credit (Age Concern, 2005d).

The majority (59 per cent) of Employment Credit claimants fell into the 50–54 age group, and were previously claiming JSA (72 per cent), while 7 per cent were previously claiming Incapacity Benefit / Severe

Disablement Allowance (Grierson, 2002). Clients have generally moved into full-time employment, although a large minority – almost one-third – have moved into part-time employment. The proportion of those moving into self-employment has been around 12 per cent.

An initial evaluation found that the main element of the programme was felt to be the Employment Credit, about which views were favourable (Atkinson *et al.*, 2000). Clients felt that it was an incentive to take low-paid work, in terms both of level and reliability of income. It also seemed to have a positive effect in terms of reducing reservation wages (the lowest wages at which a worker would be willing to accept a particular type of job). While the level of the credit was considered acceptable there were concerns that it only lasted for one year. It also tended to be most effective in geographical areas where there were low wages and low living costs. There is evidence that the ending of the Employment Credit caused financial hardship for some, something that has been considered in the design of the Working Tax Credit (Moss and Arrowsmith, 2003). Another issue was that, while the majority of clients' work histories involved higher-paid and skilled work, the majority moved into low-paid service or manual jobs which were unskilled.

A further survey (Atkinson and Dewson, 2001) found that the value of the Employment Credit was less clear-cut. Over half those surveyed stated that they would have taken a job anyway, although it had encouraged almost two thirds (63 per cent) to take a job earlier than they otherwise would have, and over two fifths (43 per cent) to stay in the job for longer than they might otherwise have done. The main factors influencing whether or not a client had entered work were that they were aged 55 or younger, female, had not had a long spell of previous unemployment, and had been convinced by the availability of the Employment Credit to take a job at a lower wage.

Few had heard of or were interested in the training grant element of the programme. Also, support provided by a personal advisor was less likely to be cited than the Employment Credit (Atkinson *et al.*, 2000). A likely factor contributing to the extremely low uptake of the New Deal 50 Plus training grant was that clients had no experience of buying training for themselves, and therefore had little knowledge about what they needed or where to access training (Atkinson and Dewson, 2001). Other factors appear to be that in-work support in terms of developing a training plan was not available, and a belief that new skills were unnecessary for the kinds of jobs being taken (Moss and Arrowsmith, 2003).

Other research has been undertaken in response to policy interest in finding out what happens to New Deal 50 Plus clients once the

Employment Credit ends (Grierson, 2002). The New Deal 50 Plus evaluation database was used to investigate clients who had completed their claim over six and 12 months previously. For the client group for whom there were 12 full months of benefits data after the expiry of their Employment Credit:

- Some 84 per cent were not claiming benefits at the 52-week stage.
- Some 77 per cent had not returned to claim benefits at any stage during the 12 months following the end of their entitlement.

There was a substantial difference in job retention rates between 'Job Centre Plus' clients aged 50 and over in general and Employment Credit claimants six months after their job starts. The latter had a higher retention rate of 84 per cent, compared to 70 per cent for the former.

Thus, if sustainability of employment is considered, New Deal 50 Plus has been a success, and the programme also appears to benefit clients in terms of improved self-confidence, increased motivation and more effective job search activities. However, a significant minority of clients feel demeaned by low pay and unskilled work (Moss and Arrowsmith, 2003).

New Deal 25 Plus and Pathways to Work

For people aged 50 to 59 who had been claiming JSA for 18 months, in 2004 the government launched a pilot study to trial mandating their participation in the New Deal 25 Plus Intensive Activity Period. Long-term unemployed job seekers aged 25 to 49 are already required to participate in this programme because it offers extensive help back into work and has a mandatory 'Gateway' element for all individuals unemployed for 18 months or more and in receipt of benefits. Job seekers aged 50 and over who had claimed JSA for 18 months could volunteer to take up this extra help and any additional options such as the Working Tax Credit. As of December 2003, the number of clients aged 50 and over who had participated in New Deal 25 Plus was 73,910, according to the scheme's database. According to the government, older workers often fail to participate because many have grown demoralised about their prospects of returning to work.

The government has also implemented Pathways to Work pilots, focusing on Incapacity Benefit claimants (Department for Work and Pensions, 2004). Main features include:

- specialist personal advisor support;
- a series of work-focused interviews during the first 12 months of a claim, when a return to employment is a more realistic option;

- greater responsibility and power for personal advisors to manage clients via easier access to government programmes, including New Deal for Disabled People and work-focused rehabilitation programmes offered jointly by Job Centre Plus and the local National Health Service;
- a financial incentive for claimants to return to employment;
- involvement of other local stakeholders, e.g. employers and General Practitioners.

The government said that it would invest almost £100 million to pilot these changes. Evaluation of the initiative included consideration of whether the approach was equally effective for all age groups. Importantly, it appears to have been rather less effective in helping people aged over 50 make the transition from being a benefit claimant (All Party Parliamentary Group on Ageing and Older People, 2006).

Further vocational training

The main training programme for the unemployed is WBLA, which is directed at those aged 25 years and over who have been claiming JSA for at least six months. WBLA aims to help unemployed people move into sustained employment and to help long-term unemployed people gain skills in areas where there are recognised skills shortages.

WBLA offers short job-focused training, longer occupational training, basic employability training and self-employment provisions (SEP). Findings of a study which aimed to investigate the use and experience of WBLA by people aged 50 and over and the factors associated with participation, achievement and successful provision were as follows (Department for Education and Employment, 2001):

- In 1999, people aged 50 and over were under-represented in WBLA compared to their share of all long-term unemployed people.
- WBLA leavers aged 50 and over were almost as likely as those aged 25–49 to achieve a qualification, 37 per cent of the former compared to 38 per cent of the latter in 1999–2000.
- Fewer older leavers found employment, 36 per cent compared to 41 per cent in 1999–2000.

Much of this difference is explained by higher proportions of longer-term unemployed and of basic employability trainees among those aged 50 and over. Furthermore, officials mentioned that they were finding it difficult to use WBLA provision to meet the needs of their older clients (Winterbotham *et al.*, 2002). This tended to be either because such clients needed intermediate/advanced skills training and provision of such courses was limited, or because the most appropriate provision was often basic employability training and this was particularly difficult to persuade older customers to access.

Partial retirement

The notion of partial or gradual retirement is now extensively debated in the UK. The government in its Pensions Green Paper stated its belief that people would prefer a gradual end to working life. However, at that time Inland Revenue rules allowed people to work and draw an occupational pension, but only if they were no longer employed by the company paying the pension. Following consultation on the simplification of the pensions tax regime (Inland Revenue, 2002), since 2006 the government has allowed schemes to offer people the opportunity to continue working for the sponsoring employer while drawing their occupational pension. Linked to this, the government has announced that it will increase the minimum benefit age from 50 to 55 by 2010. In future the concept of a normal retirement age will no longer feature in tax regulations, although some pension schemes may choose to continue with it if they find it makes sense to do so. The government is consulting with employer, employee and pensions industry organisations on how to promote best practice.

Combatting age discrimination and prejudices

Governments over the last decade have placed considerable emphasis on awareness raising among employers in particular. An important element of recent policy has been the Code of Practice on Age Diversity. Its impact, however, in terms of changing employer behaviour, appears to have been slight. Small and medium-sized enterprises have been particularly difficult to reach and the government has taken steps to address their specific needs. Non-government agencies have played a key role in working with employers in terms of tackling age barriers, although significant age-related labour market barriers remain.

Since the year 2000 the government has run Age Diversity in Recruitment Awards of Excellence and has aimed to raise awareness of the issue in specific sectors via the placement of articles in trade publications and more broadly via articles in the general business and regional press (Education and Employment Committee, 2001). A further initiative is the Age Positive website, launched in 2001, which includes a variety of features including employer case studies, advice and guidance.

Following the publishing of the European Equal Treatment Directive there was substantial debate in the UK around the form and content of legislation proscribing age discrimination. As a first step, the government consulted widely about the issues the directive raised for the UK. It presented initial proposals in 2001 (Cabinet Office, 2001). Then in the consultation document *Equality and Diversity: Age Matters*, the Department of Trade and Industry set out more detailed proposals.

The specific areas of employment covered were the age of retirement and recruitment, pay and non-pay benefits and unfair dismissal. It was stated that in exceptional circumstances, treating people differently on the grounds of age would be allowed, although employers would need to justify doing so and be required to produce supporting evidence if necessary. In July 2005 the Department of Trade and Industry produced draft regulations in light of responses to the consultation. The legislation came into force in October 2006. Interestingly, there is evidence that in the leadup, some employers were deliberately dismissing older workers (BBC News, 2006). Apparent concerns among businesses may partially result from fears that an ageing workforce might increase costs to the employer of insurance-based benefits (from an online press release by Aegon UK, no longer available). Surveys point to widespread antipathy among business to employing older workers, highlighting both the need for measures aimed at tackling workplace age discrimination and the huge challenges in overcoming it (from an online press release by Thomas Eggar, no longer available). Meanwhile, efforts are under way to widen the scope of laws on age discrimination beyond that of the workplace (BBC News, 2007).

Active labour market measures in Germany

Social Security Code III

In Germany, Social Security Code III regulates labour market measures for the unemployed. The law stresses the need to assess the individual competencies and deficits of each unemployed individual and to derive necessary measures from these. In some measures, older unemployed people are considered a specific target group. The Social Security Code requires that older people, severely disabled people, long-term unemployed, and people who want to return to their job after a period of absence must receive assistance. Furthermore, employment agencies are obliged to keep a record of how these target groups use active measures.

Regarding job placement, employment agencies provide assistance to a greater proportion of older than younger workers. In 2003, some 390,000 (15.7 per cent) out of the 2.45 million unemployed people who found a job did so with help from the Federal Employment Agencies (Bundesanstalt für Arbeit, 2004). However, the integration of older unemployed in job placement activities is rather low: workers aged 50 and over only constituted 12.1 per cent of all job placements by the employment agencies in 2003, whereas they represented 24.5 per cent of all unemployed. This lack of integration is partly due to the fact that the ratio of job placement agents to clients is low. On average, one job agent deals with 400 unemployed persons. While it is intended to achieve a ratio of one job agent to every 75 unemployed persons, this will not be achievable in the near future. Even officials from the Federal Employment Agency doubt that a better ratio than 1 to 150 will be realisable.

In the following paragraphs, a descriptive overview of the main active labour market measures that are currently in force in Germany is provided. The newly established specific measures for older workers mentioned earlier are not included because assessment is weak so far.

Further vocational training

Such measures in Germany encompass a broad range of short-, mid- and long-term training schemes via which individuals can obtain occupational knowledge and skills. They can also offer opportunities for career advancement or job changes and provide vocational qualifications. The Federal Employment Service bears the costs incurred in further training directly. These include course fees, subsistence allowances, and costs of accommodation and travel. The only precondition for this type of funding is that the employment service regards these measures as 'necessary' for job placement.

Recent labour market reforms (Hartz I and II) have curtailed the length and amount of funding for individual training measures. This has resulted in a substantial reduction of programme entrants, disproportionately affecting older workers (Winkel, 2003). This is due, in part, to the fact that the only training measures that are funded are those that are able to integrate at least 70 per cent of participants into the labour market, and that employment agencies do not actively pursue the integration of older unemployed people into public training programmes, particularly when high unemployment rates among them suggest a lack of job prospects.

Testing and short-term training measures

Since 1998, so-called 'testing and short-term training measures' have been part of the employment promotion law. These training measures are meant to test the ability and willingness of job seekers to work, and to allow for short-term training and probation periods. They are funded for a period of up to eight weeks. All registered unemployed people are eligible, even if they do not receive unemployment benefits.

Wage subsidies

Age-dependent wage subsidies are paid as 'integration' subsidies (*Eingliederungszuschüsse*) to companies that recruit job seekers aged 55 or older. As a temporary measure, the age limit was lowered to 50 until 2006. As a rule, subsidies are paid for a period of 24 months and amount to 50 per cent of all paid wages. Until 2003, for older workers these subsidies could be increased up to 70 per cent and the payment period could be extended up to 60 months. After two years of payment a reduction of 10 per cent occurs. From 2004, this subsidy was limited to 36 months.

Promotion of self-employment

Until recently, the only funding that was available for unemployed people intending to become self-employed was a so-called 'bridging allowance', implemented in 1986. It is an allowance paid voluntarily by the Federal Employment Agency and is equivalent to unemployment benefit rates, which could be obtained instead. Since 2004, there is a legal right to this kind of funding, although it does not cover any investments costs. It is paid for six months during the establishment period of a business and requires an approved business plan on behalf of the beneficiary. A new instrument for promoting self-employment was created in 2003, known as the 'me-incorporated' provisions.

Job creation measures

Job creation measures are a special kind of wage subsidy programme, since a subsidy is seldom paid to private enterprises but rather to public agencies and non-profit organisations. Projects that are promoted by the Federal Employment Agency must be in the public interest and not able to be carried out without this type of support. The aim of job creation measures is to provide temporary employment for unemployed people, in particular long-term unemployed persons, for up to one year. The subsidy pays up to 80 per cent of the normal wage. Structural adjustment measures (SAMs) are very closely related to job creation measures but preconditions for receiving these subsidies are less strict and their

Table 3.5   Selected active labour market programmes and participation rates of older workers, 2001 and 2004

Programme	Average number of participants per year, 2001 (000s)	Ratio of older unemployed, 50 and over (%)	Average number of participants per year, 2004 (000s)	Ratio of older unemployed, 50 and over (%)
	All age groups	50 and over	All age groups	50 and over
Training (long-term)	345.0	8.0	184.5	3.9
Training (short-term)	51.3	14.0	94.7	10.9
Job creation	166.5	36.6	85.7	33.1
Structural adjust- ment measures	76.5	31.9	31.5	62.2
Wage subsidies	109.4	43.9	110.5	52.1
Self- employment	43.1	11.1	154.5	13.0

Source: Bundesanstalt für Arbeit (2003, 2005).

duration is longer (up to three years), although actual payments are lower. Older unemployed workers aged 55 and over could receive wage subsidies for up to five years up to the end of 2003, but since then the duration has been limited to three years.

The effects of the respective measures for older workers are rather mixed. To summarise (see also Table 3.5):

- The most important programme, which funds long-term training, is that in which older unemployed people participate the least. Participation rates recently have dropped from an already low level of around 8 per cent to around 4 per cent;
- Short-term training measures also show low and declining participation rates among older unemployed people, declining from 14 per cent in 2001 to around 11 per cent in 2004.
- Job creation shows slightly decreasing participation, but structural adjustment measures in contrast show increasing participation rates among older unemployed people, and they were clearly over-represented in these programmes with ratios of 33 per cent and 66 per cent respectively in 2004.
- Wage subsidies also showed increasing participation rates of older unemployed people, and ranked second with a ratio of older clients of above 50 per cent in 2004.
- Promotion of self-employment again is marked by low participation rates among older unemployed people.

Table 3.6 Selected active labour market programmes and integration ratios, 2004 (%)

Programme	All age groups	50 and over
Training (long-term)	38.0	29.6
Training (short-term)	34.6	25.9
Job creation	23.0	20.3
Structural adjustment measures	37.1	37.1
Wage subsidies	68.0	61.0

Source: Bundesanstalt für Arbeit (2005).

High or low participation rates among older unemployed people within these respective programmes do not demonstrate their effectiveness in terms of labour market integration. Using so-called 'integration ratios' to analyse outcomes and effects of specific labour market programmes, the following can be concluded (see also Table 3.6):

- In almost all cases, the integration ratio of older unemployed people is lower than for the unemployed in general.
- At first glance, the most successful programmes for unemployed people aged 50 and over are wage subsidies, with an integration ratio of above 60 per cent. However, high deadweight effects need to be considered.
- Both long-term and short-term training measures show only medium integration ratios (30 per cent and 26 per cent respectively).
- Job creation measures and structural adjustment measures in particular show similar integration ratios both for older unemployed and unemployed people in general. However, it should be noted that such measures are only successful because of further support that a client can access.

Concerning individual measures, the older unemployed are still severely under-represented in short- and long-term training measures and often actively excluded from them. Effects of age-specific training provisions are limited thus far. While representation is above average with respect to wage subsidies and job creation measures, these measures are either prone to deadweight effects or lack integration into the primary labour market. Promotion of self-employment is limited even though success rates are rather high. In general, integration ratios of older unemployed workers are lower than for the unemployed in general. Job placement programmes are severely understaffed and older job seekers are often forced to give up the status of being registered unemployed. Newly established 'personal service agencies' are not geared towards the needs of older workers either.

Partial retirement

Amid the current discussion of labour market and pension policy, part-time work for older employees is regarded as an alternative to early retirement and as an instrument for encouraging more to remain in employment. Thus, the official aim of the Partial Retirement Law (Altersteilzeitgesetz, ATG), enacted in 1996, was to ease the transition from work to retirement and provide the opportunity to reduce working time for a certain period. Furthermore, its objective was to promote compensatory recruitment, particularly of unemployed workers or trainees, and thus replace older workers with both younger and unemployed workers. The main features of the Partial Retirement Law are as follows:

- Working-time reduction from full-time to half-time employment is possible from the age of 55 onwards for a period of up to ten years.
- Working time can be reduced on a daily, weekly, monthly or yearly basis if an average work time of no less than 18 hours per week is ensured.
- The reduction in employee income due to part-time work is compensated for out of unemployment insurance funds for a period of up to six years.
- To ensure at least 70 per cent of the last net salary and 90 per cent of pension insurance contributions, subsidies are provided by the Federal Labour Office.
- Preconditions for this compensation are that available part-time jobs must then be filled by unemployed persons or trainees.
- Partial retirement agreements at firm level or among the social partners must last for more than three years.

Takeup has risen significantly in recent years. In the five-year period from 1998 to 2003, the number of cases funded yearly rose from about 13,000 to more than 75,000. However, the number of funded cases does not represent the actual adoption of partial retirement in Germany. Since no state subsidies are given in the first 2.5 years of the most commonly used 'block model', these cases are not fully registered in the statistics and have to be estimated. Additionally, companies sometimes use partial retirement without claiming state subsidies and without replacing part-time work with an unemployed worker or trainee. Data available from the social security agencies show that at the end of 2003, approximately 235,000 older workers aged 55 to 65 were using partial retirement in some form or other.

Judging by these figures, it seems that the Partial Retirement Law has been a success so far. However, this does not mean that a gradual transition from work to retirement for a substantial number of older workers has been accomplished, or that their integration into the labour market has been achieved. The block model requires older employees not to change to part-time jobs but work full-time for 2.5 years and then leave for a pre-retirement sabbatical of another 2.5 years. The block model, in combination with pension insurance opportunities, which offer workers the opportunity to retire at the age of 60 if they adopt a partial retirement measure, could perhaps be more accurately described as a modified instrument for early retirement than a step towards gradual exit. Federal statistical data show that the block model is the dominant measure adopted by employers and employees. In 2002, only 11 per cent of all new entrants for funded partial retirement were for non-block models. Thus, on the whole, the Partial Retirement Law is not actually used to enhance part-time work options for older workers.

Combatting age discrimination and prejudice

Despite the existence of EU actions to prohibit discrimination in the labour market on grounds of age, Germany only came up with a draft version of a law against discrimination in autumn 2004. This belated development is partly due to the fact that the term 'age discrimination' has not found much recognition amongst politicians or academics working in the field of old-age policy (Frerichs and Naegele, 1997a, 1997b). It may also be because of the strong opposition of employers' associations to statutory measures, which are viewed as bureaucratic and ineffective and at the same time as limiting an employer's freedom of choice. By contrast, trade unions favour enacting legislation against discrimination in the labour market, although views on this are split because early retirement continues to be viewed by many as a socially acceptable means of facilitating exit from the labour market. Finally, in 2006 an age discrimination law that prohibited direct and indirect discrimination on grounds of age and in the labour market in particular was enacted.

Besides age discrimination laws, some federal initiatives exist which explicitly draw attention to the competencies and potential of the ageing workforce and thus try to counteract discriminatory behaviour and prejudice based on beliefs that older workers are less productive.

Campaign '50 Plus – They Can Do It'

In 2000, the Federal Employment Agency launched a long-term campaign called '50 Plus – They Can Do It'. This campaign was designed to foster the recruitment of older skilled unemployed workers through the ongoing job placement initiatives of regional employment agencies and to promote their reintegration into the labour market. The following aims were set out:

- to change attitudes of employers towards older workers and to combat age stereotypes;
- to increase the motivation and ability of older unemployed people to apply for new jobs;
- to improve the job placement initiatives of regional employment agencies.

Demographic change – public relations and marketing strategy

Following the completion of a research programme on 'Demography and Employment', the Federal Ministry for Education and Science launched this transfer project between 1999 and 2003 with the aim of heightening awareness of the impact of demographic change on employment. The main target groups were companies, employer and trade union groups, employment services and regional development agencies (Buck and Dworschak, 2003).

Perspectives for Germany – a strategy for sustainable development

In 2002, the federal government presented its overall strategy for sustainable development. Incorporated within it is a sub-theme called 'Potential of Older People in the Economy and Society' (Bundesregierung, 2003). Activities supported by this strategy aim to raise awareness of demographic change which could aid the economy, employment systems and societal development alike. Increased health, wealth and qualifications of older people in the workforce are viewed as key components. The strategy will be supported by pilot or so-called beacon projects run by companies which promote the employability of ageing workers and encourage lifelong learning.

Fifth Commission for Reporting on the Situation of the Elderly

In 2003, the Federal Ministry for the Family, Seniors, Women and Youth established this Commission, which focuses on the subject of 'promoting the potential of older people in the economy and in society'. In 2006, the Federal Ministry launched its report which contained wide-ranging recommendations on how to retain, promote and use the potential of older people, particularly in the labour market (Bundesministerium für Familie, Senioren, Frauen und Jugend, 2006).

New quality of work

In 2002, the federal government launched the national initiative 'New Quality of Work' (Initiative Neue Qualität der Arbeit (INQA)). This initiative is part of Europe-wide activities for more and better jobs. In 2003, as part of this initiative, a sub-group was formed to promote employment for older workers. It is intended to create a network of actors involved in occupational health promotion to foster action for older workers and to promote awareness of good practice examples.

Comparison of active labour market measures

The question arises as to whether, given the new risks for older workers – that is, the potential for prolonged spells of unemployment and fewer opportunities to take refuge in early retirement measures – the newly installed or planned active labour market measures as described take account of these risks in terms of strategy, scope, and resourcing, and/or are able to distribute the risks between advantaged and disadvantaged groups of older workers more evenly.

In the UK, New Deal 50 Plus – and in particular the Employment Credit component – New Deal 25 Plus, and Work Based Learning for Adults are measures of greater significance. Job creation, promotion of self-employment and other measures like work trials are negligible in terms of numbers of participants or virtually non-existent. In Germany, general long- and short-term training measures, wage subsidies, job creation schemes and promotion of self-employment are in place as well as selected measured targeted at the older unemployed in particular, though no integrative measure such as New Deal 50 Plus has been invented.

Job placement, as the essential and initial measure to be taken to reintegrate older unemployed into the labour market, seems to be more advanced in the UK and provides better resources – both in staff numbers and access criteria – than in Germany. It is yet to be proven if Germany can develop similar structures. There, job placement programmes can be characterised as understaffed for the purpose of addressing the target group of older workers appropriately, and job agents – encouraged to focus on the easiest to place – lack the will to address the needs of older long-term unemployed people. However, the backdrop in the UK is that besides Employment Credit (now ended) and WBLA, they have not much to offer in terms of active measures. In particular, the very long-term unemployed and Incapacity Benefit claimants are not assisted to any great extent.

Concerning further vocational training, older workers in both Germany and the UK are under-represented. In the UK, under-representation of older workers compared to their share of the long-term unemployed is reported for WBLA as well. Nevertheless, training forms a substantial part of active measures for older workers in the UK (yearly entry numbers between 10,000 and 12,000 for the unemployed aged 50 and over) and even more in Germany (yearly entry numbers until 2003 were around 50,000 for the same age group, unemployment among older workers being very much higher). It needs to be remembered, though, that in Germany the average length of training measures is presumably much higher than in the UK; funding of higher skills training is far more developed, and in general, not only the training as such is subsidised but also maintenance of the trainee. However, in the wake of recent labour market reforms (Hartz I and II) in Germany, a substantial reduction in entries to training programmes, which has disproportionately affected older workers, has taken place. In consequence, participation in Germany has fallen by almost two thirds.

Wage subsidies and in-work income support (respectively employment-conditional tax-credit schemes) are other important features when referring to active labour market measures for older workers. In the UK, until 2003 Employment Credit administered through New Deal 50 Plus, were granted. Since its launch in April 2000, about 100,000 Employment Credit starts were counted. The assessment of takeup is mixed, and it is clear that the client group is heavily skewed towards those claiming active benefits. In Germany, employer subsidies are the prevailing measure and much more funding is in place for this than has been the case for Employment Credit in the UK. In Germany the average number of participants in the year 2002 alone amounted to about 110,000, with significant increases in recent years. Additionally, it has to be borne in mind that the duration of grants is very much longer.

In principle, Germany has a very advanced instrument to promote partial retirement, and funding is ample. However, it is mainly used as an instrument of externalisation which cannot be justified to that extent, not merely because it contradicts the shift of paradigm towards integration, but because it favours the more privileged older worker. The UK has yet to overcome general obstacles to gradual retirement in the occupational pension system, though a willingness to do so exists.

The UK has relied to a high degree on publicity campaigns to promote age awareness among UK employers. Comparing the existing awareness campaigns according to their degree of intensity, it can be stated that in Germany the initiatives were slow and tentative and started late at the end of the 1990s. Announced changes in paradigm, the campaign '50 Plus – They Can Do It' and the most recent initiatives of the federal Chancellery show that demographic changes witnessed in the last few years have been acknowledged as important rather late by almost all relevant partners in Germany. In the UK, since 1993 consecutive campaigns have been undertaken and this exhibits continuity in state structures and a strong attachment to voluntarism, as could be expected from a low-interventionist liberal welfare state. This system, albeit highly advanced, does not appear to succeed fully because it is relying strongly on minimal state involvement and leaves the door open to voluntary initiatives.

Both Germany and the UK recently outlawed age discrimination in employment and vocational training. However, age discrimination and awareness campaigns can play only a supporting role when supplying more active measures. In this respect, the role these features play in the UK may be overbalanced though they fit into a strategy of low direct intervention by active measures.

Overall, the main deficit concerning active labour market measures for older unemployed people in Germany is the lack of specific targeting of those people, both in active job placement and training. Current programmatic and financial restrictions concerning vocational training further increase this deficit. Employment agencies – against a background of a lack of resources and the demand to be effective – concentrate on those who are easier to place. Older unemployed people are often omitted or even encouraged to utilise early exit provisions. Furthermore, measures offered to facilitate a transition to retirement are still in force, though not differentiated as between those actually in need and those older workers who are relatively well off. This is especially the case concerning partial retirement and structural adjustment measures. It can be argued that Germany has still to develop a comprehensive strategy for active ageing to ensure that people remain in work for longer, particularly after the age of 60, and to increase access to training for older workers in the context of lifelong learning.

In the UK, the 'joined-up' approach is stronger, at least regarding coordination structures. However, the scope of active measures seems rather limited both with regard to the types of measures – Employment Credit and intensified job placement in New Deal 50 Plus / New Deal 25

Plus – and the level and duration of funding. Meanwhile, age barriers in the labour market still appear to be a major problem, and to date, limited progress seems to have been made in tackling these. Why it has been so difficult to tackle ageism in the labour market is a matter of conjecture, though a previous long-term emphasis on early retirement, the complexity of the issue and a lack of understanding among employers, and that it is easier to use age as a proxy for performance than formal assessments when making employment decisions, may be factors.

Conclusion

A prolongation of working life requires that older people have a realistic chance of working longer. An improvement in the labour market situation will not suffice *per se*. Both incentives and provisions for individuals to work for a greater number of years and for employers to recruit and retain older workers must be improved and backed by legislative measures. However, at the same time, socially acceptable pathways into early exit need to be preserved for certain 'at risk' groups of older workers whose prospects of remaining in or re-entering employment are low, for example, workers with severe health problems and disabled workers as well as those in organisations where work pressures are high.

On the surface, strengthening of policies towards the reintegration of older workers and a reversal of early retirement has taken place in both Germany and the UK. This is embedded in a general trend of fostering active and activating strategies, which emerges against a background of labour force ageing and concerns about the future funding of pension systems, and is partly influenced by the 'moral pressure' of the Open Method of Coordination at the EU level. In Germany, with the Job-Aqtiv-Law and Hartz Reforms, profiling measures for the unemployed in general and specific support for older unemployed people have been implemented. In the wake of implementing the principle of 'support and demand', requirements for the unemployed to be active in job search and for job acceptance have been intensified. In the UK, an even more rigid placement regime has been in force for longer, whereas active policies have been implemented only since the Labour Party came to power in 1997. In this regard, New Deal 50 Plus, New Deal 25 Plus and Work Based Learning for Adults are measures of note.

The preceding descriptions and analysis of labour market policies for older unemployed people in the UK and Germany show, however, that neither the 'active' side nor the 'passive-activating' side of the labour market policy of each country is 'future-proof'. Concerning active policies it can be argued that:

- Firstly, labour market policies for this target group are not advanced enough to adequately support the need to reintegrate older claimants and, in particular, older long-term unemployed people and those on disability and sickness-related benefits. While compared to the situation at the end of the 1990s, a clear shift towards strategies and measures for older unemployed people has taken place, they remain under-represented in general labour market measures such as the promotion of vocational training, self-employment and job placement. In particular, strategies and concrete measures for improving lifelong learning and providing training for all age groups are underdeveloped.
- Secondly, even if measures are taken, they tend to benefit more advantaged and easier-to-place older unemployed people. Risk groups such as the low-skilled or disabled are not strongly represented. Even though profiling measures have been implemented to assess needs, 'creaming' and deadweight effects remain widespread.

Although new elements have been added to labour market policies in the UK, they are still grounded in what could be called 'post-Thatcherism' (Lewis, 2001). In the UK, the flat-rate system of unemployment benefits and social assistance still means that job losses immediately turn into a direct threat of undermining the financial status of those employees who are made redundant. The abolition of the Earnings-Related Supplement has put the UK in the position of being the only country in the EU in which unemployment benefits are not linked to earnings (Hassel, 2001). Due to comparatively weak worker protection and union power, many redundant older workers are more likely to be remain unemployed, or be re-employed often at lower wages, which in turn leads to more inequality/income polarisation amongst older workers (Esping-Andersen, 1999). This risk is reinforced by a lack of social protection and still insufficient active labour market measures. It is questionable whether programmes such as New Deal 50 Plus can substantially minimise these risks.

In Germany unemployment risks and income poverty among older unemployed people may increase. Exclusion from employment of older workers – and women in general – is still the dominant feature so far. In the past, this has been cushioned by early retirement and employment insurance, and in consequence poverty among the elderly is largely absent. However, the restriction of the formerly generous early retirement option and severe curtailments in respect of unemployment benefit and unemployment assistance have taken away some of the most important adjustment mechanisms to respond to worsening economic conditions, and have therefore created a more pressurised situation. In the policy debate, the emphasis on social security has weakened. Substantial pressure has been put on employment protection for older workers as well. Whether, in future, active employment programmes will be a suitable replacement is questionable. Retrenchment of early exit has accordingly not led to an adequate expansion of the funding of active labour market policies. On the contrary, older unemployed people are increasingly at risk of being neglected in training programmes, and targeted programmes for this group and general improvements in the job placement structure have yet to prove whether they really do work. Critical voices emphasise that the true target of the new anti-pre-retirement policy in Germany is the lowering of pension costs and of funding for active labour market policies, not the promoting of employment opportunities for older workers.

However, and despite a broad array of critics of the recent reforms seeing a radical shift or at least a convergence of Germany to a (neo-)liberal welfare state regime, this opinion cannot be maintained when looking at structural indicators – such as level and duration of unemployment benefit in Germany compared to the UK – and structural reforms of institutional arrangements, such as the still high cooperation of employers' associations and trade unions in the reformed Federal Employment Agency. In general, for Germany path-dependency has to be stressed, which despite changes of government has kept the alignment of the welfare state regime rather stable. The same holds true for the UK the other way round. Despite a more socially inclusive stance and the implementation of target-group-orientated active labour market programmes in recent years, a substantial shift towards a conservative-corporatist regime cannot be observed.

It is not by accident that Germany adopted age awareness campaigns only recently and that age discrimination legislation is still in its infancy, given the more regulatory employment system. Nor is it surprising that in the UK, funding of job creation programmes and a broad application of training and learning measures has not taken place yet, given the low-interventionist character of labour market policies. And depending on the path each country is following, the social risks involved for older workers differ. Whereas in Germany high unemployment amongst older workers persists, but still on a better socially and financially secured level, in the UK entry to and exit from the labour market is more flexible but creates far greater income and career risks for the older workers involved.

Therefore, the main conclusion of this comparison of Germany and the UK is that despite similar challenges and at first glance similar answers – more activation on the one hand, more curtailing of early exit on the other – a rather distinct mix of, and underlying bases for, policy measures exist. The pressures of labour force and population ageing do not induce the same answers to the same problem, but are dealt with on the basis of the respective welfare regimes, economic structures and normative values in each country.

However, it should finally be noted that this chapter has been restricted to an analysis of the strategies behind and an assessment of the effects of public labour market policies for older workers in the UK and Germany against a background of different welfare regimes. But labour market policy as such is just one important policy field when attempting to improve the employment prospects of older workers. It needs to be embedded in an integrated policy approach together with pension and taxation policies, policies for promoting health and education, and policies to enhance work–life balance. Furthermore, employers play a central role when it comes to maintaining the employability of the ageing worker.

4

The Silver Economy – Purchasing Power and the Quest for Quality of Life

Wolfgang Potratz, Tobias Gross and Josef Hilbert

Introduction

It is a well known fact that the population is getting older. However, in the public and academic discussions, it is mainly the negative consequences which are being debated. Despite obvious threats to society there are further issues related to demographic change which must not be overlooked. While people are getting older they also tend to be richer than previous generations. A British commentator has put this in a nutshell:

> Given the increasing number of older people – many with money to spend – and the diminishing pool of younger ones, any company that ignores the purchasing power of older consumers is not going to be in business for long. (Skapinker, 2005)

Today's 50-plus generation is strongly influenced by the economic upswing after the Second World War. Even more so than in Germany the British system is characterised by market liberalism with powerful and experienced consumers. Not only is this generation shaped by economic freedom but it is also self-confident and critical with respect to demand. Moreover today's baby boomers aspire to staying vital, healthy and active. Contrary to general perceptions they have desires which are commonly attributed to younger people, such as for regular gym sessions and adventure holidays. Consequently products and services that promise to keep people young are highly appreciated. Mature consumers have three main characteristics in demand terms (Rizal, 2002): firstly, they are highly heterogeneous, depending on their financial, health and living situation; secondly, they are more selective and critical than younger age groups, meaning that more efforts are necessary to satisfy their needs; and thirdly, although older people tend to buy products that are similar to those that younger people prefer, the range of products in the shopping basket is smaller. However, when speculating on how the pre-retirement generation is likely to use its purchasing power, which can be quite considerable, much depends on how far market sectors are prepared to innovate in order to meet demands.

Given this, the following contribution will focus on aspects of age-specific consumption and quality of life from a comparative perspective. Is there sufficient substance, in terms of both income and demand, to drive forward a silver economy? What do people who are about to leave or have left the 'productive sphere' spend their money on, why would they like to spend it, and what do they actually get for it? The chapter takes a comparative approach, assessing, first, the demographic situation and income levels of older people in both countries. The conclusion is that, although there are differences in income levels between the two countries, it is time to review traditional perceptions of older people as poor and needy. Against this background, we will investigate what research and surveys tell us about preferred fields of consumption and where there are supply gaps to be filled. The message here is that instead of saving, older people want to enjoy life. Finally, we briefly review actual policies to push and support the constitution of a silver economy, benefiting not only the old but the economy as a whole. Due to limited comparability of available data, examples will be taken from each country respectively. Despite these limitations, it will become clear that significant similarities exist in substance but clear differences in approach, which call for more encompassing comparative research and strategic deliberation.

Demographics: change of perception

Europe is ageing but at different speeds. In comparison to the EU-25 (as of 2005), the UK holds an intermediate position with a forecast old-age-dependency ratio of 33.2 per cent in 2025. Germany is clearly above average with a ratio of 39.3 per cent (Eurostat, 2005). As Table 4.1 outlines more explicitly, for the different age groups this trend is expected to continue during the first half of this century. At the same time the working-age population will decline as the baby boomers move into retirement. Still, 'ageing is not destiny but convention, and it is not biology which makes us a retiree but culture' (Kruse, 2003). However, only since the late 1990s has the 'deficit approach' to the ageing of

Table 4.1 Population numbers and demographic change by age group, 1966–2050

Age	Country	1966 (Number)	1980 (Number)	2000 (Number)	2005 (Number)	2050 (Number)	Change 1966–2000 (%)	Change 2005–2050 (%)
50–54	Germany*	4,522,493	4,706,638	4,796,696	5,583,041	4,622,991	+6.1	−17.2
	UK	3,526,750	3,207,385	4,010,904	3,686,536	3,929,245	+13.7	+6.6
55–59	Germany*	5,115,638	4,423,149	5,150,374	4,503,962	4,804,103	+0.7	+6.7
	UK	3,470,650	3,379,636	3,262,712	3,902,612	4,150,029	−6.0	+6.3
60–64	Germany*	4,770,304	2,887,251	5,624,031	5,204,837	5,073,838	+17.9	−2.5
	UK	3,094,800	2,833,129	2,890,487	3,072,905	4,117,482	−6.6	+34.0
65–69	Germany*	3,846,087	3,979,358	4,084,930	5,193,252	4,819,583	+6.2	−7.2
	UK	2,456,950	2,836,593	2,603,963	2,707,996	3,796,188	+6.0	+40.2
70–74	Germany*	2,735,830	3,613,849	3,578,108	3,611,538	4,295,220	+30.8	+18.9
	UK	1,834,100	2,378,303	2,339,798	2,338,084	3,345,795	+27.6	+43.1
76–79	Germany*	1,806,148	2,570,847	2,848,852	3,006,133	4,254,942	+57.7	+41.5
	UK	1,277,500	1,675,107	2,007,355	1,939,068	3,404,365	+57.1	+75.6
80+	Germany*	1,382,226	2,050,457	3,010,745	3,546,736	10,162,855	+117.8	+186.5
	UK	1,143,700	1,519,850	2,365,343	2,631,919	6,577,002	+106.8	+149.9

* Including ex-GDR from 1991

Source: Eurostat Database (2006); authors' own calculations.

society been challenged by a 'competence approach' which builds on the individual, societal and economic potential of older citizens (Baltes and Mayer, 1999; Kruse and Schmitt, 2005). The 'winds of change' had already blown for a while from various directions: gerontology and related sciences had shown that improved care, modern nutrition, exercise, and societal integration support healthy and independent living; the social and economic sciences had shown that improvement of working conditions for older employees fostered enterprises' innovativeness and that where seniors' demand was sufficiently served new jobs were created; and, finally, politicians had become aware that the activation of seniors' potential better helped to contain welfare costs than assistance programmes provided later on (Hilbert and Cirkel, 2005).

This shift of paradigm is important to note because it opens up a new and more holistic view on the ageing of society and, consequently, also opens up new coping strategies. This new view became manifest both in the UK and Germany in government reports on the ageing of society (UK: Department for Work and Pensions (DWP), 2005; Germany: Bundesministerium für Familie, Senioren, Frauen und Jugend (BMFSFJ), 2006a, 2006b). In Germany, the *Fifth Report on Older Persons* (BMFSFJ, 2006b), written by an expert group of gerontologists, social scientists and economists, builds on five guiding principles:

1. age as a motor for innovation;
2. rights to and duties of lifelong learning;
3. prevention (social, educational, medical), understood as competence building for independent and responsible life;
4. sustainability and intergenerational solidarity;
5. joint responsibility for social cohesion.

Very similarly the UK's *Opportunity Age: Meeting the Challenges of Ageing in the 21st Century* report (DWP, 2005) reflects the changed role of older people in society and sets three political priorities:

1. to achieve higher employment rates overall and greater flexibility for over-50s in continuing careers, managing any health conditions and combining work with family (and other) commitments;
2. to enable older people to play a full and active role in society, with an adequate income and decent housing;
3. to allow us all to keep independence and control over our lives as we grow older, even if we are constrained by the health problems which can occur in old age (DWP, 2005, p. xiii).

The common thread running through both these reports and the resulting strategies is that of enabling and activating (through adequate support) older people to take matters into their own hands for as long as possible. It takes account of the fact that the seniors of today are, on average, richer than earlier generations, better educated and informed, more experienced in handling new technologies and more sophisticated in their demands for quality of products and services. Retirement and old age are looked upon as a new phase of life, offering chances for new experiences, to resume forgotten activities, to engage socially. Thus the *Fifth Altenbericht* (BMFSFJ, 2006b) devotes two long chapters to social participation, engagement , networking and volunteering by older people. Thus retreat from working life does not mean retreat from society and societal life but rather the opposite: explicit hedonistic attitudes are gaining ground and are being acted out. In both countries research has revealed that:

> Although people like the idea of being able to leave a bequest, most do not think that older people should be careful with their money just so that they have something to bequeath. The majority of older people said they will enjoy life rather than worry about inheritance. (Soule *et al.*, 2005, p. 78; see also, UK: Rowlingson and McKay, 2005; Germany: Hilbert and Naegele, 2001; comparative: Grumbach *et al.*, 2002)

These observations display a significant reversal in attitudes within just one decade: until the 1970s saving and inheriting were regarded as values *per se*. However, during the 1990s only half of the 60–69 age group (of which again two thirds admitted to a certain leaning towards hedonistic positions) shared this view. Yet this group makes up the most important sub-group of silver consumers (GfK Wirtschaftstrend-forschung, 2002). The changing attitudes of older people themselves, and changing societal, economic and political perceptions of older people as actors capable of organising their own life within a given social context, have come together in the idea and concept of a silver economy. It assumes that there is a market for products and services which improve the quality of life in and until old age. The viability of this idea is supported by a further glance at demographic change: not only does the growing proportion of older people in the population change the population structure, but also the structure of the older population changes – older people not only become more numerous, but also more people become older (Table 4.1). In other words, as the age phase extends, the age structure of the older age group broadens and with it social backgrounds, consumption patterns, needs and preferences can be differentiated.

Despite obvious differences between Germany and the UK the data provided in Table 4.1 show very clearly that consumers are ageing, and it does not need elaborate research to predict that the kind of structure of need and demand will change too. At least as important a message of these data is that the speed of ageing only leaves a short time span to reorganise and restructure supply in order to meet this changing, silver demand and to build a firm basis to make use of economies of scale and scope alike.

The income situation compared: money is not the bottleneck

For a long time age was closely linked to poverty. With deteriorating health and declining ability to work older people had to rely on support from relatives and the social system which was just marginally above the subsistence level. However, with steady increases in wealth after the Second World War social security systems were upgraded and personal savings grew. Today 'one of the greatest success stories of the economic and social entity that makes up Germany is, that at the end of the 20th century it is possible, by and large, to assume a satisfactory situation regarding income in old age' (Hilbert and Naegele, 2001, p. 75). On average, Germany's seniors today can dispose of material resources like no generation before. The average (weighted) income of persons over 65 currently approximates 90 per cent of German average income (Cirkel *et al.*, 2004) (Table 4.2).

Broadly the picture in Britain looks similar. Following Lloyd George's Old Age Pensions Act of 1908, the National Insurance Act of 1946 laid the foundations for a modern pension system. Consequently pensioners' net income from a variety of sources grew in real terms between 1979 and 1996 by 64 per cent (Balchin and Bullen, 2005). Over the ensuing eight years pensioners' average income grew by 21 per cent while earnings rose by 13 per cent over the same period. According to a survey by National Statistics this was a result both of increased state benefits and of the increased coverage and value of private pensions. In 2003/04 the average pensioner-couple household had a gross income of £415 (€600) per week. Single pensioners received on average £209 (€300) per week (Table 4.3).

Despite these positive trends, there are significant variations in the level and the structure of incomes (Alterssicherung in Deutschland, 2003; Schmähl, 2005b; SHARE, 2005), and variations are more distinct

Table 4.2 Net weekly income of seniors (65+) by type of household / marital status, Germany, 1992–2003

Type of household / marital status	Income (€/week)				Change (%)	
	1992	1995	1999	2003	1995–99	1999–2003
All pensioner units	1,207	1,350	1,451	1,610	+7.5	+11.0
Couples	1,695	1,871	1,958	2,159	+4.6	+10.3
Single men	1,210	1,330	1,356	1,476	+2.0	+8.8
Single women, of which:	928	1,037	1,100	1,171	+6.1	+6.5
Widows	936	1,050	1,122	1,197	+6.9	+6.7
Divorced	801	885	897	992	+1.4	+10.6
Unmarried	946	1,033	1,083	1,145	+4.8	+5.7

Source: Alterssicherung in Deutschland (2003)

Table 4.3 Average gross weekly income of pensioner units, UK, 1994/95–2003/04

Type of household / marital status	Income (£/week)					
	1994/95	1996/97	2000/01	2001/02	2002/03	2003/04
All pensioner units	225	235	269	277	280	291
Pensioner couples	326	344	387	399	394	415
Single pensioners	163	169	197	197	205	209

Source: Department for Work and Pensions. Data for all years retrieved from www.dwp.gov.uk/asd/asdb/pensioners_income.asp.

within than between the two countries. With regard to the level, the main dividing lines in Germany are former occupational status and sex. Old-age pensions in Germany are based on (compulsory) statutory social insurance contributions in relation to (eligible) income, and consequently pensions vary with the sum of contributions during working life. Since female labour market participation in Germany is still relatively low, single women, on the whole, find themselves at the lower end of the income hierarchy. Structurally, statutory pensions to a growing extent only constitute a part of old-age income. In addition to the statutory pension scheme, enterprises (either single or branches-wide), as well as the public authorities, provide supplementary pension schemes to their employees. So, together with private insurance contracts, private wealth and inheritance, there is a quite differentiated structure of old-age income. It is basically this combination of income resources which makes for (growing) differences in old-age income.

Similarly in the UK, sources of income are diversified and private pensions have become more important in older age. Approximately 81 per cent of single pensioners and 93 per cent of couples have income additional to state benefits either in the form of occupational pensions which are provided by a former employer, or personal pensions (Soule *et al.*, 2005, p. 69). However, research conducted by the Pensions Commission found that:

> The distribution of current pensioner income is highly unequal, not only because of disparities in lifetime earnings, but also because of the wide dispersion of private pension provision. Moreover the historic state system has left major gaps in provision for people who have had interrupted paid working lives and caring responsibilities, in particular women. (Pensions Commission, 2005)

Despite rising incomes, when comparing these data to the then EU-15 the UK recorded significant risks of old-age poverty (Goodman *et al.*, 2003; Council of the European Union, 2004; Banks *et al.*, 2005). The UK ranked fifth in terms of low income for people aged over 65. This is significantly different from Germany which has been ranked in twelfth position (Soule *et al.*, 2005, p. 74). Nevertheless, reforms of the British pension system such as the Pension Credit might lead to additional income in the future and reduce the risk of poverty.

Although old-age poverty has consistently gone down in Germany, some 12 per cent of pensioner households still live on or below the poverty line, as against a general poverty risk of 13.5 per cent (Bundesministerium für Gesundheit und Soziale Sicherung (BMGS), 2005; BMFSFJ, 2006b, p. 198;). Generally, old-age income reflects the overall spread of income, and likewise shares in general income forecasts: as job careers become more and more disjointed and inconsistent, so will old-age incomes. Thus, within a calculable time frame the income situation of older people may change considerably (Motel-Klingebiel *et al.*, 2004).

It is also worth noting that significant regional disparities in pensioners' incomes can be seen in the two countries. In Germany, the East–West divide leads to income differences of 25 per cent for single households and 19 per cent for couples. In the UK there are not only clear differences between Wales, Scotland and England, but also within

England with the better-off living mainly in London and the South (Balchin and Bullen, 2005, p. 24).

Although the pension schemes are structured differently, surprisingly the income spread is larger within countries than between countries. So, two trends can be perceived in both countries. On the one hand, compared to previous generations today's older people are relatively well-off. Nevertheless, there remains a considerable risk of old-age poverty for certain groups and in different regions. Any policy that aims to address the economic potential and risks brought about by demographic change must, therefore, take those income differences into account, and so must enterprises. However, given the demographic situation, changes in macroeconomic demand will, *ceteris paribus*, be unavoidable. This means that products and services need to be provided adequately for lower- *and* higher-income groups.

Consumption compared: goodbye to heirs

'Enjoy life rather than worry about inheritance' – this, in a nutshell, is the message from recent consumer surveys of the 50-plus population. As has been shown in the section above, seniors command increasingly substantial purchasing power to underpin their demand. A study by the German Institute for Economic Research (DIW) calculated that 60-plus households in 2003 spent €308bn, which is almost a third of total German private consumption (BMFSFJ, 2005b). Confirming earlier studies (GfK Wirtschaftstrendforschung, 2002; see also Reichert and Born, 2003; Augurzky and Neumann, 2005), research found that silver consumers are discerning, demanding and quality-orientated. They expect reliable information, competent service and advice, and they expect 'intelligent', functional products and services which provide comfort and quality of life. Seniors present themselves as relatively free-spending: while all private households spend roughly 75 per cent of their income on consumption, households with retired persons aged 65–75 spend approximately 84 per cent. Also for the less old, aged 60–65, as well as for those over 75, the consumption ratio is still around 80 per cent, i.e. above the overall average (Kott and Krebs, 2005) (Figure 4.1). It is no surprise that the level of spending is directly related to the level of income. However, the inclination to spend money and to consume is a common feature of all classes. The general explanation is that consumption behaviour will be approximately constant even after entering pension age, as people do not change their lifestyle so easily (Tesch-Römer, 2004; Metz and Underwood, 2005).

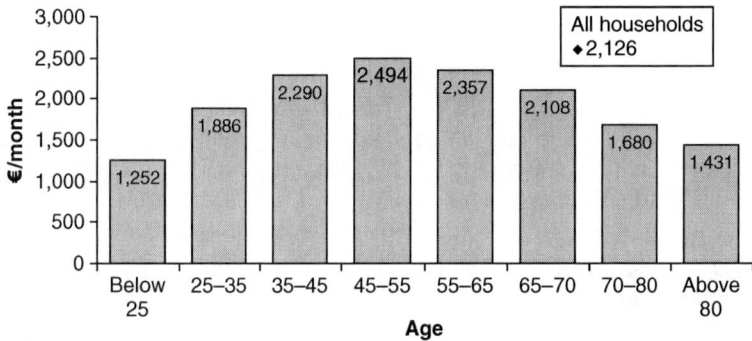

Source: Statistisches Bundesamt (2004b), p. 35.

Figure 4.1 Consumption of private households by age group, Germany, 2003

Generally as people approach retirement age they have more time available for leisure activities. As highlighted above the majority of people are willing to enjoy life rather than saving for future generations. For example, the average age of customers buying a Harley-Davidson or a Porsche is 52 (*The Economist*, 2002). This could be the result of two factors. Firstly, preferences and attitudes have changed drastically; and secondly, a change in the consumption patterns of the older population who have more financial resources and leisure time available has taken place. Clearly, there is a positive correlation between health, economic situation and lifestyle.

When looking at the main spending categories, the British example outlined in Table 4.4 highlights the fact that an increasing share of expenditure is devoted to food, housing and fuel, familiar correlates of low income, as well as recreation, culture and health.

Where the household reference person was aged 50–59, 10 per cent of spending in 2003/04 was on food, compared with 15 per cent where the household reference person was aged 80 or over (Soule *et al.*, 2005, p. 67). Nevertheless, a relatively large share was spent on recreation and culture and on miscellaneous goods and services. To a certain extent, this lower level of spending on food is reflected by smaller household size due to children leaving home. The very old are more likely to be in single-person households. Younger households are, in contrast, more likely to have people in work (Soule *et al.*, 2005, p. 79).

These patterns can be observed in both Germany and the UK, as illustrated in Figure 4.2, which outlines the development of basic demands and needs over time.

Table 4.4   Household expenditure as a percentage of total expenditure by age of household reference person, UK, 2003/04

Commodity	Age			
	50–59 (%)	60–69 (%)	70–79 (%)	80+ (%)
Food & non-alcoholic drinks	10	12	15	15
Alcoholic drinks, tobacco & narcotics	3	3	3	2
Clothing & footwear	4	5	4	3
Housing, fuel & power	9	9	12	15
Household goods & services	9	8	9	7
Health	1	2	3	2
Transport	15	14	10	12
Communication	2	2	3	3
Recreation & culture	14	16	14	13
Education	1	0	0	0
Restaurants & hotels	7	7	6	5
Miscellaneous goods & services	8	7	9	9
Other expenditure items	15	12	13	13

Source: Soule *et al.* (2005), p. 79.

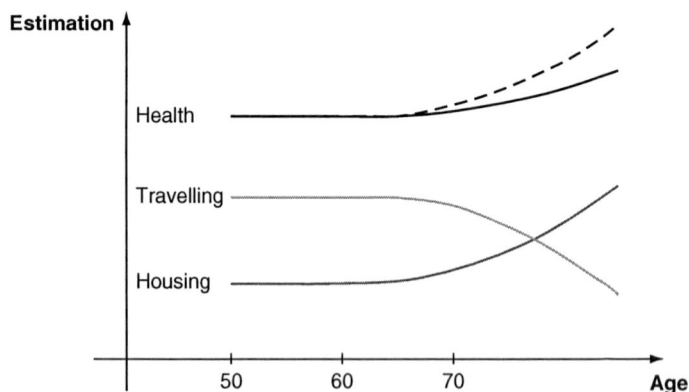

Figure 4.2   Silver economy – ideal-type demand curves over age

In the following sub-sections, therefore, these three core spending categories – housing, travelling, health – which basically constitute quality of life (not only) in old age are outlined and briefly compared. Age-related consumption patterns are for the time being clearly 'under-researched' (Börsch-Supan, 2005, p. 4), both in market research and in social analysis. Therefore, a deeper analytical approach and the necessary differentiations would go beyond this chapter (for Germany this has been done in the framework of the *Fifth Altenbericht* (Fachinger, 2004).

Housing

In modern and individualised societies, independent lifestyles are valued highly. Older people have described independence as the ability to make their own decisions and not to be a burden to others (Audit Commission, 2004a). Appropriate housing conditions are a necessary prerequisite of independence. Consequently goods and services that help people to remain in their homes as long as possible are highly appreciated. Demand in this field has gone up consistently, the more so as time spent at home increases with age. For obvious reasons, the large majority of older people wish to stay at home for as long as possible and avoid moving into residential accommodation. This is mirrored in the expenses for their homes and the adaptation of homes as needs progress. On average, therefore, housing takes the biggest share of the budget. This is only to be expected since the living situation as well as access to social and commercial infrastructure usually also determines the degree of social integration and increases the scope of independent living.

On average German seniors spend roughly 30 per cent of their income on housing (Reichert and Born, 2003; Fachinger, 2004), with tenants' budgets being stretched more than those of owners. Included in this share are expenditures on rebuilding and reconstructing homes, which, however, are not limited to barrier-free redesign. Rather, the driving motives often are new individual concepts of how to live after children have left the home, and plans for the next phase of life.

In 1989 the British government published the White Paper *Caring for People*, which sought to increase older people's independence by providing services and technology (Audit Commission, 2004b). Several subsequent White and Green Papers have reconfirmed that goal. Only recently the *Opportunity Age* report acknowledged that:

> A large part of the nation's housing stock was built with little regard for the needs of older people, who can, as a result, find themselves excluded from living an active life for no other reason than their house is ill-suited to their needs. In England, around 45 per cent of people over 85 in 2001 lived in housing that did not meet decency standards. (Department for Work and Pensions, 2005, p. 35)

In order to achieve this goal, the government aims to build an integrated home visiting service which is supposed to offer full care, benefits and housing checkups. As Figure   points out, such household services relate to an important aspect of quality of life in age, the more so as family and neighbourhood support is declining. This applies not only to care services, but primarily to professional help in and around the home, attractive delivery services and, crucially, safety (emergency hotlines and so on). Roughly 75 per cent of German seniors and almost as many of their British contemporaries would be willing to pay for such services – if they were on offer.

However, national as well as state data, e.g. for North Rhine-Westphalia, the most densely populated German state, show that although demand and purchasing power are present, these services are still significantly underdeveloped. There are many pilot projects run by housing companies and welfare organisations on all variations of 'intelligent' and 'serviced' housing, but so far none has really experienced a broad breakthrough. Similarly in the UK, it seems that there remain several obstacles that hinder supply meeting demand (see 'Excursus', below). Hence incentives for product and service development would be beneficial from a welfare development point of view as well as with regard to pushing a sustainable silver market.

Excursus: assistive technologies in the UK

In the UK, an important contribution to independent living is seen in so-called 'assistive technology' (AT). AT can be defined rather broadly as 'any item, piece of equipment, product or system that is used to increase, maintain or improve the functional capabilities and independence of people with cognitive, physical or communication difficulties' (Audit Commission, 2004b, p. 3). Obviously, there is a variety of examples ranging from walking-sticks to tele-health-monitoring facilities. Calculations have shown that, compared to nursing and sheltered housing, AT significantly reduces the costs of care. For example, staffing costs, which represent the highest share of total nursing costs, will be reduced considerably when people are able to live more or less independently at home. However, a crucial question is whether the supply of AT products is sufficient to meet demand or whether market failures hinder the efficient allocation of resources. At the current stage it seems that several obstacles remain. These include 'the volume of change faced by public services, the lack of consumer pressure in the AT "market" and problems with finding the initial investment money needed to get started' (Audit Commission, 2004b, p. 28). As a consequence of the age gap, consumers

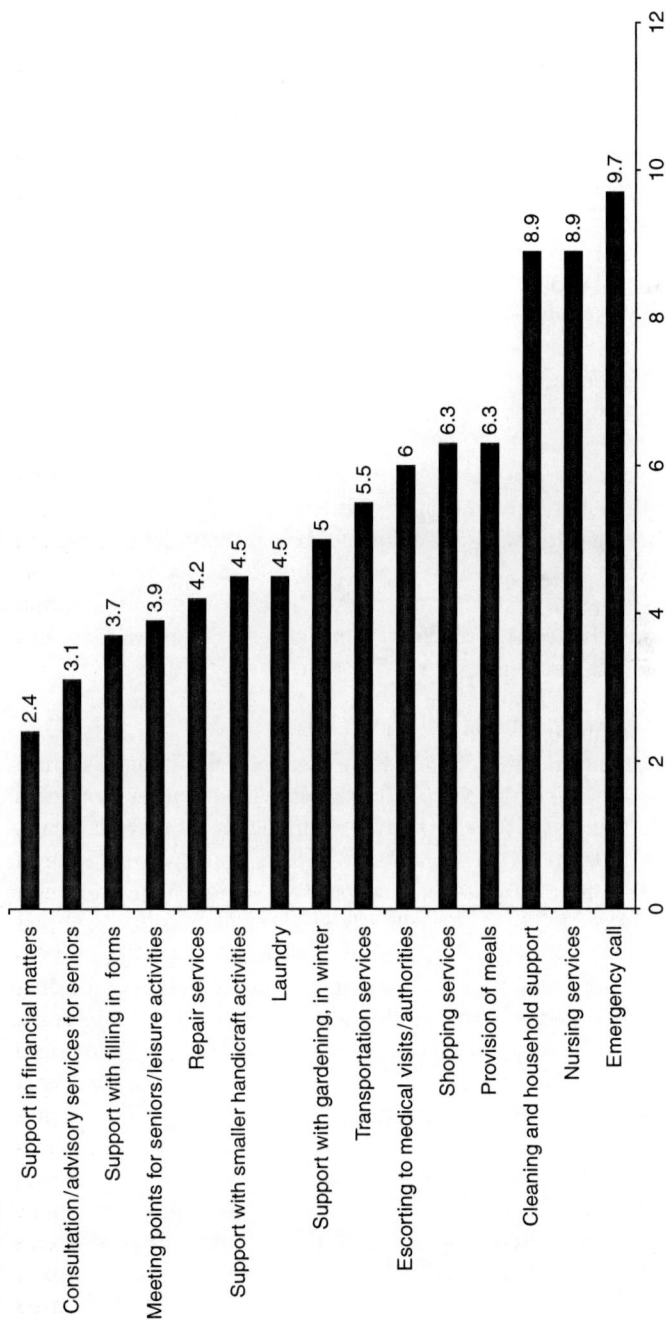

Support in financial matters — 2.4
Consultation/advisory services for seniors — 3.1
Support with filling in forms — 3.7
Meeting points for seniors/leisure activities — 3.9
Repair services — 4.2
Support with smaller handicraft activities — 4.5
Laundry — 4.5
Support with gardening, in winter — 5
Transportation services — 5.5
Escorting to medical visits/authorities — 6
Shopping services — 6.3
Provision of meals — 6.3
Cleaning and household support — 8.9
Nursing services — 8.9
Emergency call — 9.7

Source: GfK Wirtschaftstrendforschung (2002), 15 most demanded services.

Figure 4.3   Seniors' demand for household services in Germany (projected customers in millions)

of AT products are relatively distant from producers in comparison to the normal goods market.

Thus, for the time being, the actual use of housing and household services is marked by an income gap. To make total demand and supply meet, an extension of services at middle and lower price levels is clearly necessary. Particularly in middle- and lower-income classes, currently much service work is carried out in a do-it-yourself way and through neighbourhood networks. In the long term, however, this will not be a reliable and sustainable perspective, as family bonds loosen (not least due to required labour market flexibility) and also neighbours grow old. At the same time, household services hold considerable job potential, which is far beyond most traditional industrial chains and which is also related to a wide range of qualifications.

Recalling the financial situation of older people, it is highly likely that many people would be willing to pay for such products in order to increase their independence. Thus, there is no need for public services to be the sole providers. It is, however, vital that a critical mass of products is achieved in order to exploit economies of scale and bring prices down. Clearly, as the demographic situation changes the market for manufacturers and retailers selling directly to consumers will increase (Audit Commission, 2004b, p. 15) in both countries.

Excursus: financial services and real estate in Germany

An attractive field of high potential is specialised real estate. To a growing degree banks and financial institutions see this sector as a growth market, and the run on the German (semi-)public housing estate companies in 2005 can partly be explained by this newly discovered interest in the financial and purchasing powers of the silver generation (see Hypovereinsbank, 2001a; Deutsche Bank Research, 2003a, 2005; Allianz Group, 2004). The expected age structure will boost the demand for smaller but comfortable flats and also for adaptation. The market for adaptation of flats and residences only is estimated at €70–100bn for the coming decade. In respect of stationary care a demand of an additional 3,000 homes for about 300,000 persons until 2020 is forecast; for 2050 a demand for 1.5 million places is expected. Financial studies do not give monetary volumes for this market, since there are too many uncertainties about the development of pension and care schemes. An indicator of expected profitability and return on investment is perhaps the observation that one-day workshops and conferences on the issue of 'living in age' charge fees of well over €1,000 and are still fully booked. In any case, public schemes are not expected to ever be capable of covering upcoming costs, so that there will be a large volume left to be met by private expenses. This will certainly lead to a large and differentiated private market.

Overall, the silver housing market is highly promising economically and seems set to grow relative to the housing market as a whole. The market for accompanying housing and household services and assistive technologies is also expected to expand. What must not be overlooked, however, is that housing and housing environments are essential to social integration. As age progresses this becomes the focal point of quality of life. And yet, the market must be seen as being in its initial phase and sometimes the motor is still prone to splutter.

Health

'Healthy ageing' regularly ranks highest on almost all surveys related to consumption, quality of life, satisfaction, life perspectives and related issues. Yet 'healthy ageing' means not only nursing and care in case of sickness, but also 'well-being', i.e. the extent to which one is able to participate in societal life (in their clear and straightforward way the Finns call their respective programmes 'FinnWell' and 'Well-being and Health'). In this case 'health' also stands as a cipher for social integration. Accordingly, the 'demographic challenge' to the health system, in terms of growing shares of expenditure *on* older people, is only one aspect of the debate. It is also important to focus on additional health expenditure *by* older people.

Although statutory health expenditure on pensioners is constrained, quite a number of them are prepared to spend additional money on constitutional drugs, vitamins, homoeopathic drugs and treatments, 'alternative' medicines and so on (Key Note, 2005, and see Table 4.5), and on devices for home health control (blood pressure, etc.), which makes the respective pharmaceutical and medical equipment industries beneficiaries of demographic change (Deutsche Bank Research, 2003). Similarly cures and treatments in health resorts are very popular, even when additional co-payments are required. In all, an increasing awareness of body, health and well-being across all age sub-groups is gaining ground and is fed by a growing willingness to pay (BMFSFJ, 2006b, p. 247). According to the consumption statistics for 2003, all age groups spent between €75 and €80 per month on additional health expenditures (out-of-pocket payments), while those aged between 55 and 70 spent roughly €110 per month (Statistisches Bundesamt, 2004, p. 76, Table A18). It must be noted, however, that in official statistics

Table 4.5 Complaints suffered in past 12 months and use of alternative health care practitioner or health food shop/herbalist by age, UK, 2003/04 (% of respondents)

Complaints in the past 12 months	Age						
	15–19 (%)	20–24 (%)	25–34 (%)	35–44 (%)	45–54 (%)	55–64 (%)	65+ (%)
Cough	71.5	73.1	65.9	59.5	50.3	46.9	38.4
Cold	76.5	76.1	74.3	68.4	59.8	53.2	37.5
Headache	76.8	74.8	75.9	71.3	61.7	49.4	29.9
Hay fever	25.6	27.1	23.0	21.1	16.4	10.9	8.6
Catarrh	13.9	15.5	16.2	17.9	20.4	20.3	19.3
Indigestion	23.5	27.1	30.0	32.0	31.4	32.7	30.4
Insomnia	15.0	23.0	20.3	22.9	22.9	21.9	18.9
Arthritis	4.6	5.7	4.4	6.9	17.7	29.5	38.7
Skin complaint	31.6	24.5	22.0	19.2	18.7	17.3	14.2

Source: Key Note (2005).

on income and consumption these developments are only partially covered, since products and services are aggregated differently and in different categories. As a result, these figures tend to underestimate true levels of expenditure.

In Germany today, members of statutory health insurance schemes finance roughly 7 per cent of all health expenses by out-of-pocket payments, and up to 2015 a marked increase in private expenditures is predicted. This will also drive a change in demand and demand behaviour, as 'patients' will gain in competence and change from being recipients of assistance to self-reliant 'customers'. Many of these additional expenses go into what is regarded as prevention or simply 'medical wellness'. In particular, markets for technical health, rehabilitation and gerontological devices are expected to rise, as today's 60-plus-ers move into older age and make their way through the various phases of age-related prevention, health and care. Besides technical devices, pharmaceutical products and functional foods are increasingly prominent. These claim to explore the full potential of prevention and rehabilitation and to further improve quality of life; 60 per cent of seniors look upon 'nutraceuticals' and dietary supplements as components of healthy food and nutrition (GfK Wirtschaftstrendforschung, 2002, p. 118).

One case in point is wellness and fitness centres. Being close to health, sports and exercise facilities is seen as a strategy to prevent disease and slow down the physical and mental ageing processes. Staying fit and active ranks high and is closely associated with the abovementioned body awareness (GfK Wirtschaftstrendforschung, 2002, p. 215). As well as attending courses many older people possess bicycles, ergometers of some kind and other fitness-related devices, and spend copiously on outfits. Until recently the 'fitness industry' had neglected this layer of customers and experienced a period of slowdown; however, when the industry had responded to the changing demand structure with age-specific fitness centres and respective services, growth rates improved significantly, and a field for new businesses opened up. Older people remain a reliable group of customers: some 65 per cent of the 55–69 age group do sports (more or less regularly), and even in the 70-plus age group the rate is still 36 per cent (BMFSFJ, 2006b, p. 250).

Excursus: fitness, physiotherapy and health – no strange mixture

'Oase' in Bochum (North Rhine-Westphalia) is a 'high-end' fitness cen-tre, but prefers to be looked on as a 'health and sports club'. Under this label 'Oase' does, of course, offer high-end facilities, but the dis-tinctive feature is the provision of additional accompanying services aimed at their 40-plus customers. Services and marketing are driven by a dedicated holistic health orientation rather than by lifestyle. Cus-tomers go through a thorough health check before exercise or training courses, vital parameters are documented and controlled, encompassing anamnesis interviews are conducted to find out about current or ear-lier impairments, and age- and health-specific programmes are designed accordingly. Customers are looked after by qualified and constantly trained staff, and in addition there is a 'medical connection team' who give medical advice and if necessary contact physicians or therapists. The fee is about €60 per month and membership is growing.

As prevention strategies are gaining ground in the design of health policies as well as in individual behaviour and estimation, markets open up which often combine medically induced therapies, preventive measures and 'simple' medical wellness. However, the fuzziness in def-initions as between 'medically induced' provisions, which are usually paid for out of health insurance, and 'wellness' activities, paid in part and depending on contractual provisions by insurances and/or out of the older person's pocket, is mirrored in statistics. Due to the methods of aggregation, expenditures are systematically undervalued. However, building on case studies and observations in the field, the demand for health and related products and services can reasonably be expected to expand in the years to come.

Travelling

Mobility, along with cultural activities, ranks high on the seniors' agenda. For a quarter of the 50-plus-ers travelling is almost an elementary need, and consequently expenditure has grown steadily. On average a month's income per year is spent on holidays or other travel. As is well known, Germans love to travel, and they do not basically change this attitude as they get older. In North Rhine-Westphalia, which is the largest but not the richest German state, 55-plus households on average spend more than €2,200 per year on travel, which is 8.2 per cent of net household income (Reichert and Born, 2003, p. 37). Silver travellers tend to be characterised by the following attributes, they:

- travel farther and more often than younger people;
- use public transport facilities;
- prefer package tours;
- stay in hotels;
- travel in groups and appreciate contact with others, communication and animation;
- change to short journeys as they grow old and handicapped;
- thoroughly evaluate the attraction of the destination and the possibilities/ease of getting there;
- highly value availability of medical support and general assistance at the destination;
- critically assess barrier-free design of locations and establishments (Institut für Freizeitwirtschaft, 2003; Cirkel *et al.*, 2004).

With this consumer behaviour in mind, travel agencies and destinations can look to a bright future, as the propensity to travel is increasing despite and against all economic trends. These prospects are also driven by a differentiation of journeys: the yearly 'summer holiday' becomes shorter and additional second and third breaks are undertaken with different objectives, e.g. to enjoy cultural events, to do city tours or to follow educational activities ('edutainment').

Similarly to Germany, in the UK travel for all people increased substantially from 1985, but remarkably so for elderly people. Journeys made by elderly men and women increased between 1985/86 and 1996/98 by 12.2 per cent and 18.8 per cent respectively (Banister and Bowling, 2004). Surveys have shown that there is a clear positive correlation between quality of life and transport (*ibid.*). Those older people who had access either to their own car or alternatively proper public transport rated their living quality as good or very good. However, there is a tendency for older people to become less inclined to travel abroad. Instead they are more often choosing regions in the UK as holiday destinations (Soule *et al.*, 2005, p. 86). This means that there is rather good potential for tourism development in the British regions.

Quite a number of travel agencies have focused on this specific target group of silver travellers, and with increasing demand, supply is slowly growing, too. For example in the UK the SAGA group started its business with low-cost off-peak holidays for older people. Today the company focuses exclusively on the provision of high-quality, value-for-money services for people aged 50 and over. However, the intensity of travel is closely related to education and income levels, as is demand for culture and education, e.g. the takeup of (Open) University studies for seniors, and 'edutainment' offers, be it in relation to painting, cooking or health, in culturally inspiring environments.

A related phenomenon is health tourism, which is growing more strongly than general tourism and which, as Table 4.6 displays, is driven to a great extent by 50-plus-ers. Their expectation is often to recover and keep healthy and do this in a stimulating, entertaining and meaningful way. Statistically, assuming a uniform distribution, the 50-plus age group would contribute roughly €1.9bn to the turnover of the health tourism branch of industry up to 2010. This is more than 50 per cent of the projected growth (Cirkel *et al.*, 2004; also see Illing, 2002). Translated into jobs, for Germany this would mean 48,000 jobs within this period – given an adequate and attractive development and design of offers.

Synthesising these developments and trends and specifically targeting their customers aged over 50, TUI, the largest European travel group,

Table 4.6 Market share of health- and wellness-related travel by age, Germany, 2002–04

Age	Population (million)	Fitness (%)	Wellness (%)	Health (%)	Spa (%)
14–29	12.7	4.0	2.7	2.6	0.9
30–39	11.4	4.0	3.4	3.7	2.2
40–59	20.7	4.3	5.0	7.5	4.5
60+	19.9	2.4	3.8	15.2	14.5
All 14+	64.7	3.6	3.9	8.3	6.5

Source: Lohmann and Winkler (2005), p. 19.

has invented the 'TUI Club Elan'. The club concept focuses on the interests and preferences of 50-plus customers: communication and collective experiences, fitness and wellness programmes, cardiovascular regeneration training, internet courses, and language or dancing courses. Hotels are selected from the higher categories, with a convenient public transport infrastructure, leisure facilities and medical care. Turnover realised by this concept grows more than proportionally, and more than 80 per cent of participants have joined programmes at least twice. The number of locations has been extended (Pricewaterhouse Coopers, 2006).

Quality of life – needs for (re-)design and innovation and political support?

The economic attraction of ageing results from the interplay of three basic factors:

- *Demography*: not only do the numbers and the proportion of older people increase, but in contrast to earlier generations many of them live longer in good health and actively participate in societal life.
- *Purchasing power*: again in contrast to earlier generations today's pensioners are materially well equipped and they command considerable shares of overall purchasing power.
- *Images of age*: today's pensioners see themselves as 'best agers', they feel younger than earlier 'old' generations and they display more hedonistic attitudes, i.e. they would rather spend their money than hand it on.

Seniors increasingly regard themselves as active within society, serene, curious as to what life still has in store, and open to new technologies. They fit into the metaphor of the psychologist Charlotte Bühler: 'While biologically we are going down, biographically we are looking up'. As a result, German 'best agers' represent almost a third of overall private consumption, of everything from sports cars to geriatric devices. Qualitatively challenging demand has put both manufacturing and commercial enterprises under pressure to develop and design adequate products and services. Moreover, as market research has shown, the dividing line is less the socio-demographic category 'age' and more the functional categories of quality and usefulness. Thus, one can speak of senior consumers as motors of innovation.

However, it has taken some time for policy makers to incorporate this aspect of demographic change into their thinking. Only recently has the UK government's agenda moved from focusing on supporting the minority of older people with the highest levels of health and care needs to including the maintenance and promotion of independence for many more older people. With the *Opportunity Age* report (Department for Work and Pensions, 2005) the agenda widened further, aiming to increase quality of life for everyone in older age. As outlined above, the strategy explicitly builds on enabling older people to play a full and active role in society. It aims to give older people independence and control over their own lives. It includes proposals on cultural change, tackling inequalities, employment, active ageing as part of the community, independence and control, and public services for older people.

However, there are shortcomings with regard to developing a silver market. Although it is recognised that income and wealth have increased considerably, there is no mention in the report of whether there is a sufficient supply of private goods and services. Also, analyses of the potential macro-economic benefits of the activities of silver consumers are missing. Nevertheless, there is clearly a paradigm shift visible in the sense that it is acknowledged that people over the age of 50 are far from being a 'drain on society' (Department for Work and Pensions, 2005, p. xvi). Although the financial resources associated with this strategy and the likely impact remain uncertain, with its cross-departmental and multi-policy perspective the strategy represents a further step away from the fragmentary approach of policy makers noticeable up to recent times.

Slowly this change of paradigm is also being realised by commercial companies. In the UK several marketing agencies have already started focusing on the silver market and there is clearly a rising awareness of the purchasing power of the '50-plus generation' (see, for example, such initiatives as Diametric, HeadlightVision, Millennium Direct, Senioragency, and 20 plus 30 (use of interactive technologies in marketing to the over-50s). This is due not least to influential lobby organisations such as Age Concern England, which published a study focusing on the older consumer (Metz and Underwood, 2005). And yet, 'a survey of 45,000 over-50s published in April [2004] by specialist marketing agency Millennium ... discovered that 86% felt ignored by the marketing industry, and 70% felt patronised by advertising' (Harkin and Huber, 2004, p. 44). Research has shown that, despite the private-sector acknowledgement of older consumers' attractiveness, there is still a lack of specific programmes to target this group. Rather, older people are for the sake of simplicity perceived as a segment of a wider market (Ahmad, 2002). Although comprehensive analyses of companies' marketing behaviour are beyond the scope of this chapter, academic evidence suggests that much more can be done in order to make use of the economic potential of older people. And very often it turns out that it does not need completely new products but only more intelligent design and marketing.

Conclusion

Contrary to the UK's ageing strategy, in Germany the *Fifth Altenbericht* has dropped the emphasis on 'welfare' policies and focuses more directly on the social and economic 'asset' of age and how this resource can be fed back into the economic cycle. How this can be facilitated had been successfully demonstrated by the North Rhine-Westphalian Silver Economy Initiative (*Landesinitiative Seniorenwirtschaft*) (Ministerium für Gesundheit, Soziales, Frauen und Familie des Landes Nordrhein-Westfalen, 2003). Launched in 2002, it serves as a blueprint for a number of 'quick second innovators' in other German states and regions (North Rhine-Westphalia has also pushed a European network of regions (Silver Economy Network of European Regions, Sen@er) dedicated to promoting the development and marketing of innovative products and services aimed at serving the demand from senior citizens, thereby contributing to regional development and job creation. Results of economic and social research had prepared the ground for a strategy that was built in essence around arguing (with the help of research), persuading, activating and enabling. The 'big players' in the field, welfare organisations, seniors' organisations, public authorities, trade associations and companies, were invited to discuss and design projects which they subsequently and cooperatively realised. These 'pilot projects' gave rise to more and different activities, self-organised and managed by groups of actors. Companies in all trades, whether associated with the initiative or not, have taken up the issue of ageing in their strategies, launched projects, redesigned goods and services and brought them to market. In this way an emerging silver market has gained momentum and is about to develop self-sustaining dynamics, resulting in a forecast 100,000 net additional jobs by 2010.

A question is whether such a strategy could also be realised in the UK. As has been highlighted above, income is even more widely spread between social groups in the UK than in Germany, with few at the high end and (too) many at the lower end. So critics may contend that, notwithstanding the North Rhine-Westphalia example, on the whole a silver economy will remain a niche market for the happy few. Nevertheless, economic theory predicts that it is market size and economies of scale which make goods and services generally available and affordable to a wider range of consumers. This holds particularly for new and innovative products and services. In order to make supply and demand meet, further development of this 'emerging' market is clearly necessary. And since we know very well which products and services are being demanded, it will probably need only a little political effort and support, mostly in the form of reducing information asymmetries between consumers and producers. Policy makers and associations can provide support for companies by raising awareness and setting incentives. Clearly this would not only be to the benefit of older people but also to the society as a whole, since macro-economic demand-multiplier processes are likely to trigger economic growth and (additional) employment. While it is obvious that a silver economy cannot provide solutions to all aspects of demographic change, it can nevertheless help to mitigate some adverse effects attached to ageing, leaving a brighter prospect for the future.

Acknowledgements

The authors gratefully acknowledge the helpful criticism and support of their colleagues Elke Dahlbeck (IAT) and Martin Ferry (EPRC). However, all shortcomings remain the responsibility of the authors.

5
Gender and Ageing – The Role of Social Networks

Ingrid Eyers and Gertrud Backes

Historically, in both the United Kingdom (UK) and Germany, it is social policy that has defined the onset of 'old age' for men and women. In the UK the chronological transition into 'old age' is defined by 'pensionable age' and in everyday language the term 'pensioner' is a synonym for 'old person' (Bytheway, 1997). This would be the same in German where *'Rentner'* or *'Pensionaer'* would be the relevant synonym. In both countries being a pensioner implies a lifestyle that is free of the constraints imposed by the necessity to earn money and where time is available in abundance. In reality the lifestyle of older men and women in the UK and Germany is influenced by social policies that for example determine socio-economic status or relate to health care. The range of policies that impact on an older person's life are widely discussed in other chapters in this book. The aim of this chapter is to compare and reflect on the lives of older men and women in the UK and Germany and consider how over the life course changing policies influence their present lives and enable the accumulation of social capital.

This chapter aims to bring together social theory, social policy and practice. Based on exemplars from each country the life course of two women and their social networks provides the background required to raise issues that are key to their lives as they approach their 80th birthdays. Within this context, the role of a variety of relevant policies will be outlined and the changing characteristics of men and women as pensioners and the importance of their social networks will be considered.

Background

The present-day longevity of men and women living in the UK and Germany raises the issue of adequate quality of the 'gained years'.

Above all this is demonstrated in the characteristics of the objective and subjective dimensions of the life situation, the material situation, activities, health status, housing and social contacts. Women in particular live longer and are affected more strongly by social and health problems as they grow older (Peace, 1986; Arber and Ginn, 1991; Kohli and Künemund, 2000; Höpflinger, 2002; Backes, 2003, 2004; Clemens and Naegele, 2004). The integration of social networks is gaining outstanding importance in this context (Künemund and Hollstein, 2000; Wagner and Wolf, 2001; Arber *et al.*, 2003).

In relation to the so-called critical life events this is especially true (Filipp, 1981), i.e. basic changes of a positive or negative kind, such as changes in the workplace or redundancy, changes in the life situation, changes in partnership (such as marriage, moving in together, separation/divorce, and living apart together), changes in the family situation (such as the birth or death of a child, migration or loss of friends). Such basic changes also characterise old age. Here, there are typical requirements for adjustment regarding losses as well as for making new beginnings in the middle age of life. These situations can arise when the children leave home; during the person's and/or their partner's transition into retirement; when suffering a decline in health; when care is required; in the case of loss of the partner; when the possibilities to stay living in the familiar environment become restricted or in the case of changes of housing (Fooken, 1980; Backes, 1983, 1993, 2002; Perrig-Chiello and Höpflinger, 2001). It is in these instances that the strength and importance of social networks established over the life course become highly relevant and where gender differences are noticeable in both the UK and Germany.

Social networks and social capital

To explore the meaning of social networks it is suitable to refer to Pierre Bourdieu's (1983, p. 190), term 'social capital'. This highlights the character of resources and emphasises the idea that 'the social capital is the totality of the actual and potential resources which are associated with the possession of a lasting net of more or less institutionalised connections of mutual knowing and appreciating each other, or in other terms, it is a matter of resources which are based on the affiliation in a group' (*ibid.*).

The causes of social differences and inequalities, according to Pierre Bourdieu (1983, p. 190), are:

• the social capital, the connections and networks;
• the economic funds in the traditional sense;

- the cultural capital of education and titles;
- the symbolic capital of prestige, honour, and approval.

The amount of social capital a person has acquired is dependent on the extent and mobility of their social network with its connections. Furthermore it is dependent on the economic, cultural, and symbolic capital of those persons who can be reached by this. The total capital of a network member awards them a certain degree of credit. On this basis social capital serves as a source for transactions and power (Haug, 1997). In the following we consider how women and men create their social capital and how this impacts on the lives of old people living in the UK and Germany.

Gendered networks

It is predominantly women who maintain contacts within the family and neighbourhood (Finch and Mason, 1993; Jerrome, 1993; Jeffrey, 1997; Davidson *et al.*, 2003). Throughout their lives women are active as so-called 'emotional labourers', above all in the closer or wider circle of the family, friends and acquaintances. Amongst their neighbours and in the community, old women are integrated in social networks from which they can also benefit when assistance in everyday life is required. How far this networking is sufficient to bring long-lasting assistance or even the provision of care is however questionable. On the other hand, men even in later life may well remarry or find a partner with whom they cohabit and can count on in the case of illness, or need of help or care. Because of their previous, mostly continuous labour men are more likely to have good links outside the circle of family and partnership. Potentially men are also able to fall back on close and binding family networks in the case of immediate problems in everyday life.

It is to be anticipated that husbands blend into their wife's network which is biographical and rooted in the family, neighbourhood and past employment. Such social networks contribute considerably to the well-being of both men and women in all stages of later life (Cooper *et al.*, 1999; Hoff, 2003; Fiori *et al.*, 2006). As Fiori *et al.* (2006) conclude from their research, the well-being of older people may be more strongly influenced by friendships than family.

A further factor that requires consideration is loneliness. In research undertaken by Victor *et al.* (2004), it was identified that only 7 per cent of the participants reported that they were always/often lonely. Whilst gender is linked to living alone, chronic medical conditions and widowhood, the experience of emotional or physical loneliness is not specific to men or women but old age is a risk factor (Wenger *et al.*, 1996; Fees *et al.*, 1999). However, research undertaken by de Jong Gierveld (1999) indicates that older men living alone are at risk of experiencing isolation. On the basis that women live longer than men it is more likely that the oldest old women experience loneliness. Wenger and Burholt (2003) have identified that loneliness is a dynamic process and is a secondary response to acute or chronic changes in circumstances. Such an acute reaction is to be expected when bereavement is experienced. However, such a reaction to widowhood for both men and women is gradually overcome (Victor *et al.*, 2004). The support network surrounding an older person living in the UK or Germany will contribute to the successful process of adapting to any change in circumstances. This exemplifies the importance of social networks, especially in later life. However, as Hoff (2003) points out, the network of the oldest old does decrease in size.

Gender and the division of labour

A gendered division of labour, for the majority of old women today, means that their primary role was and sometimes continues to be that of a wife and mother. Professional labour on the part of women did not need to be undertaken in order to secure their own existence – at least according to the norm. It was rather intended to happen only in case of emergency, when the husband was made redundant or his income was not high enough. On the one hand this 'female normal biography' should have meant security, which was not at all always the case. On the other hand it meant commitment to a particular form of life and material existence. The risks of the 'old' or 'traditional' way were primarily related to family commitment and financial dependency, resulting in their existence being justified only in this private family area and in the corresponding social networks. For old women this endangered their material position but also impacted on their socio-psychological situation because it was derived from and linked to the husband's work and social status outside the family.

Thus studies of old women are constantly indicating that they lead more contented, healthy and even-tempered lives when they are also part of social networks outside of their home and maintain their areas of social contact and the ability to participate in other activities. Focusing exclusively on the family can lead to poverty in old age and can also harm their health, psychological well-being and, above all, restrict

their social networking. Put briefly, qualified and continuous professional labour is considered as the best 'gero-prophylaxis' (Lehr, 1977,
1978, 1982; Szinovacz, 1982; Clemens, 1992).

This disadvantage to women is to some extent reduced when there is
an extended traditional family that compensates for the lack of social
contact outside of the family.

Potentially, men dispose of more forms of social integration procured
by their professional labour, and in old age they are more likely to have
the chance to include themselves once more in social networks outside of their home and their family. Higher social, economical, cultural,
and symbolic capital, developed during their previous life, are essential
resources here. However, as the work undertaken by Arber *et al.* (2003)
indicates, gendered roles and relationships are changing.

The experience of growing old for women can be seen to be more
contradictory, more ambivalent and socially more problematic than that
for men. This implies not only more and different impairments, but also
different, more multiplex and subjective forms of coping. This is also
true for the development of social capital and the inclusion in social
networks throughout their life course, up to old age. On the one hand,
women are seen as 'communication labourers'; they maintain family
and neighbourhood relations, and support the partner in maintaining
his professional contacts. On the other hand, in old age they remain
alone, without direct private networks, and have to cope with assistance
and care provided by strangers.

Social engagement and maintaining autonomy are acknowledged as
enhancing older people's quality of life (Gibson, 2001; World Health
Organisation, 2002; Bond and Corner, 2004; Victor *et al.*, 2004), and
older people strive to maintain their autonomy and independence with
the support and guidance of their social networks. For an older person
in both the UK and Germany, maintaining such networks is linked to
personal factors such as health, family status, and financial resources,
and to external environmental factors such as neighbourhood, transport
systems and communication systems. Gender and the life course will
have influenced the personal factors that are developed and established
by each individual, whilst the external factors are linked to the 'lived
environment' encountered in each country. Social policies are related to
both personal and external factors as they influence, for example, the
education, health and wealth of an individual and the infrastructure
of a country. It is in the everyday life of older people that the impact
of social policies on social networks can be observed. Based on the
framework of the biography and present social networks experienced

by two women of the same age, one living in the UK and the other in Germany, the personal and external factors relating to networks will be discussed next.

Comparative exemplars

The two women in the following exemplars present very different life courses which raise a wide range of issues and indicate the changes taking place in the typology of old people in the UK and Germany. They also provide an insight into the creation and maintenance of their social capital. The key commonality in the exemplars is gender and age. In order to maintain anonymity the two women have been given the pseudonyms Marjorie and Renate. Both have given their informed consent for their life stories to be used as exemplars in this book and chapter.

Prophylactic public health policies have contributed towards the successful longevity being experienced by Western society and Marjorie and Renate are exemplars of the results of these policies. Their life course has taken them through World War II and the rebuilding of both the UK and Germany. They brought healthy children into the world and any major health problems they have encountered in the course of their lives have not been related to major infectious diseases. For previous generations this could not be taken for granted and both are acutely aware of how their longevity is related to public health policies which they continue to experience, for example through their annual 'flu jabs' and regular health checks.

Marjorie and Renate are both approaching their 80th birthday; both have two children aged around 50 and have teenage grandchildren. At different stages of their lives both women have experienced major health problems such as breast cancer. Throughout most of their adult lives both were in full-time employment and are entitled to their own occupational pension which provides them with a relatively comfortable lifestyle. In their late 60s both moved considerable distances away from the cities they had lived and worked in for over 25 years. They each now live in semi-rural towns in close proximity to a relative. Marjorie in the UK is owner-occupier of a bungalow, whilst Renate in Germany lives in rented accommodation. At the end of World War II they both met their husbands. However, whilst Marjorie enjoyed celebrating her golden wedding anniversary a few years ago, Renate has been divorced for almost 40 years and since her children left home has been living alone.

Marjorie's network

Family

Retirement planning for Marjorie and her husband had always included plans to be near to at least one of their children. The move from the North to the South of England, just over twelve years ago at the time of writing, means Marjorie now lives within eight miles of her son and daughter-in-law and in closer proximity to two of her husband's siblings who live in the South East. Family bonds were strengthened by this move and the family network is extended to include nephews, nieces and their children. This is maintained and supported by annual family gatherings that are not determined only by weddings and funerals. The network builds on the relationships that Marjorie, as a single child, developed over her life course within the family she married into.

However, Marjorie is physically separated from her daughter and grandchildren who had been living in the United States for almost twenty years at the time of writing. The reduction in the cost of air travel over that period years enabled Marjorie to see her daughter and her family on an annual basis to maintain both physical and emotional contact. To further maintain emotional contact Marjorie is dependent on communication systems which now range from telephone calls to emails. Maintaining contact over a long distance has been enabled and facilitated by the reduction in cost of telecommunications and the technological developments encouraged and supported by government. Mobile phones and text messaging are not technological developments that have been taken up by the family, and mobile phones are seen to be purely for emergency use when out of the house. The interest in maintaining contact with the family in the US and within the UK has encouraged Marjorie to keep up to date with the latest technological developments in computers, where broadband internet access facilitates emails with digital-photograph attachments. This has enabled her to take part in her grandchildren's interests and development in between visits. Telephone conversations however are the preferred mode of communication as the actual sound of a voice is the closest form of physical contact that can be achieved in such circumstances.

The fact that her son lives in close proximity means that in times of need he is at hand to provide moral support and as-and-when-required assistance in everyday life activities that range from hanging pictures on the wall to chauffeuring. Knowing he is close by provides reassurance. For Marjorie the social network mutually developed with her husband over the years is key to her everyday life. However, building on his past working life her husband continues to use his management skills and takes on the role of 'managing director and procurement officer' of their everyday activities. He shops, plans and organises their activities and where possible he also supervises activities such as gardening and cooking at meal times.

Friends

The peer group of friends, who have encouraged the use of computers, was based on Marjorie's lifelong interest in music and the church. Joining the choir and becoming an active member of the church community with its activities provides a stable structure to her everyday life. This is extended by her interest in gardening and flower arranging which brings her into contact with a further network of friends. The bungalow is surrounded by similar properties. Consequently the neighbours are pensioners from a similar middle-class income group with backgrounds in varying professions but all originally from different parts of the UK.

The skills to use the latest computer technology have been encouraged and developed through this peer group of friends that Marjorie has developed since her move to the South of England. The urge to use the computer as a form of communication and the role it could play in easing the stress of the 'weekly shop' at the supermarket as well as the extended contact with the family have resulted in the installation of a computer in the bungalow. Being part of the digital world means that if wanted and needed Marjorie also has access to the latest services that the UK government is now making available online. For example, information about health and social services is readily available on the internet and Marjorie would not need to leave the house to gather information. However, once the available information has been collated it is to be expected that it will be discussed amongst family and friends. Ultimately, the experiences and knowledge of Marjorie's social network will contribute considerably to the decision-making process. This could be related to the choice of hospital and consultant for surgical procedures or the choice of a domiciliary care service or care home.

The social network is already operating as a 'mobility support system' and the group of friends take each other to and from choir practice, theatre visits, hospital appointments, and visits to friends at home or in hospital. This social network operates on a very informal reciprocal basis and functions around physical abilities, disabilities and availability.

At present Marjorie does not need to use public transport as she and her husband still drive. However, medical and other physical constraints could one day curtail their mobility. As her son and daughter-in-law are in full-time employment, Marjorie and her husband would at times be dependent on assistance from their social network or public transport.

The only service provided by the state and used by Marjorie is the National Health Service and, though reticent, she does currently visit her local doctor's surgery regularly to monitor her health and she has her annual 'flu jab'. However, as she progresses into her life as an octogenarian, changes in her circumstances are to be expected and the family and social networks will be put to the test before services set in place by health and social care policies are called upon.

Renate's network

Family

When Renate moved from Central Germany to Northern Germany, it meant returning to her 'home town' where she rents an apartment within a mile of her brother, who is ten years younger, and his wife. This is what she had always intended to do when she retired. Geographically, this also meant she was halfway between each of her two children. She is still approximately a nine-hour car journey away from each child. The closest relative from the next generation is a nephew who lives a five-hour drive away. The close relationship with this nephew builds on the years where she and her sister, both as single parents, pooled their resources and did their best to jointly raise their children. Renate's sister is no longer alive and her brother is her closest living relative from her generation. Should an acute situation arise, then as 'backup' to her brother, Renate's nephew, who is a young *Pensionaer* (56), would be able to be there as soon as possible to assist and help until one of her children arrived. This family is dispersed throughout Europe with its roots in a rural town in Northern Germany. Contact within the family is maintained through regular phone calls and visits. In contrast to Marjorie, Renate does not have a partner with whom she can discuss, arrange and undertake everyday activities.

Despite her owning a computer and a mobile phone neither form part of Renate's communication system with the family. However, the computer is used regularly to play games such as Mahjong that keep her mind agile. Like Marjorie, Renate wants to hear the voices of her nearest and dearest and text messaging is not one of her skills. For Renate the mobile phone is also seen as part of a safety-net system when out of the house. Unlike with Marjorie none of her friends are interested in using computers to gain access to the internet either to find information or to receive emails.

Friends

Whilst this might make it sound as if Renate is isolated, she has in fact re-entered and extended the circle of friends she left when she moved away to marry nearly sixty years ago. These contacts are rooted in her life course and had been loosely maintained through sporadic visits over the years and intensified by reunions linked to school and confirmation in the Lutheran Church. The strength of the childhood friendships is astonishing and provides a strong and effective network. This close-knit circle of friends consists mainly of women now living on their own. The husband of only one of her closest school friends is still alive. One of the school friends never married but whilst working full time also supported her sister who has four children and also worked full time. A relative newcomer to the circle also never married and moved to the town when she retired to be near her two cousins and where the roots of her family are also based. Three of the group never moved away from the town and are also the second or third generation of their family to live in the same community. Most of their children are dispersed throughout Germany yet these friends all have a sense of belonging and security living in this town. Whatever the problem, if between them they do not know the answer, they do know how to obtain information and guidance.

Renate fears dementia, yet feels confident that if she were to become confused, the neighbours all know her and would guide her back home and call her brother. The couple living in the apartment above her are in their early fifties, have known Renate's family for over forty years, and are very caring, supportive and helpful. Should Renate need continuing care at home, her sister-in law knows the people from the Lutheran Church and Red Cross who jointly manage the local home care services. The doctors she goes to see regularly also know her well and she knows the local hospital from being a patient within the last ten years and visiting friends, family and acquaintances who have been hospitalised. It is also the hospital that she knows from her childhood. Should circumstances be such that the move to a care home had to be made, the two local care homes are also known to her, the family and her friends. Some of the staff may well be known to her, and amongst the residents there may well be people she remembers from her youth.

Although Renate has a car and drives locally, for long-distance travel the train station is within walking distance, as are the shops and all other amenities. There is an interesting cultural programme with a high standard of musical and theatrical performances. The local open-air swimming pool in summer and indoor pool in winter are comfortably accessible. Active ageing is taking place within this circle of friends who meet up to go on walks together, will meet for a morning swim in summer, or just meet to enjoy cooking a meal that they would not consider worthwhile cooking for one person, or meet for afternoon coffee and cake followed by a game of cards.

As Renate progresses into her life as an octogenarian, changes in her circumstances are anticipated. However, despite the scattered nature of the family, it is to be expected that the changes she may encounter have already been discussed. The options available are known to all, and family and friends would pull together to ensure she receives the best possible support whatever the situation.

Marjorie and Renate present two very different stories that highlight a range of issues encountered by women approaching their 80th birthday at the time of writing, and indicating the changes to be encountered in the next cohort of the oldest old persons. However, both are part of a network that is a blend of family and friends and have accumulated social capital that extends beyond the family. Marjorie's network consists predominantly of married couples. Within her family, a brother-in-law had been caring for his wife for many years and has been a widower for three years. His social network is based on his past occupation and his visits to his only son who lives abroad. However, his academic status, interests in music and literature, and financial security have enable him to develop social networks in the city his son and daughter-in-law live in. Marjorie and her husband share the same social and family network that has been jointly created and is rooted in their professional backgrounds and lifelong interests. They each follow their individual interests within this network but these activities are not gender-specific. For example, Marjorie enjoys singing in a mixed choir, while her husband enjoys playing Bridge in mixed company. In Renate's network the men are mainly her relatives and as she herself points out, the men in her extended network of acquaintances within her own generation were amongst the last to be called up to the fight in World War II, and many did not return.

Men and their social networks

Gaining an understanding of networking from a male perspective is challenging, especially as gender-related literature in the UK and Germany to date focuses mainly on women in later life, with a strong emphasis on their socio-economic status and their health and social care needs when living alone. Men are seen as privileged and advantaged and therefore not perceived as representing a necessary area for research.

Consequently there is an under-representation of research relating to old men (Calasanti, 2003). This indicates that old men who are socio-economically disadvantaged, never married, are widowed or divorced represent a significant minority that is overlooked in both the UK and Germany. As Marjorie and Renate's networks indicate, as part of a married couple old men appear to be active within social networks. However for men living alone, this in not necessarily the case.

As described by Arber *et al.* (2003), men can be expected to experience the transition from being a breadwinner to becoming a pensioner as a loss of power and masculinity as they move from centre stage in the workforce to not even on the sidelines but totally off stage once they are no longer actors in the workforce. For men, adapting to change also involves the creation of social networks and these can be expected to be based on contacts maintained with former colleagues who are also now pensioners, and possible membership in sports and social clubs. Data from a survey conducted in Germany by Hoff (2003) provide an insight into the importance of social networks and their influence on the quality of life for men. The findings indicate that male networks develop throughout the life course. They are key both to the subjectively experienced quality of life and to the objective aspects of men's lives. Joint leisure activities and experiences combined with mutual support are the main characteristics of these relationships. Whilst the support network contributes to the improvement of the objective aspects of men's lives their emotional relationships are embedded in a circle of people who provide them with an identity and purpose in life. Intergenerational relationships, especially between parents and children, as well as relations within one generation, for instance between brothers and sisters, are the most stable connections men maintain during their life (Hoff, 2003).

Davidson *et al.* (2003) point out that men's lives in old age reflect their marital history. Yet their social worlds and behaviours cannot be evaluated only on the basis of their present relationship with their partner. Successfully establishing and maintaining social networks is not only based on an individual's gender but also relates to their personality. Extroverted old people of either gender are more likely to establish friendships and make social contacts than those who are introverted (Krause *et al.*, 1990). How successfully men living alone in the UK or Germany integrate within their own families can be expected to depend on the relationships they have with their sisters or daughters. Widowed men appear to be cautious about how they interact with older single women and do not want their actions to be misinterpreted as a signal to commence an intimate relationship. Their success in maintaining social networks will to some extent be dependent on their personality and will reflect how they have established such networks throughout their life course. However, the findings from research undertaken by Davidson *et al.* (2003) in the UK indicate that men who have never been married perceived themselves to be different, appear to be 'self-contained' and not to seek intimacy.

The intergenerational relationships of old men at the beginning of this millennium reflect their post-World War II need to focus on earning a living. Those who have children had little time to build a meaningful relationship with them. For divorced older men, a key issue was that they often had only sporadic contact with their children (Davidson *et al.*, 2003). However, as further described by Davidson *et al.*, developing good relationships with grandchildren is important to many older men and consequently this can be seen to be part of their social networks.

For married men, adapting to their changed lifestyle provides the potential to build networks that are, as in the case of Marjorie and her husband, based on both joint and individual interests. Within this relationship the man retains his power and masculinity by taking on the role of 'managing director' of their joint lives. A man living alone because he never married, is divorced or widowed, presents a different picture. However, there is a need to consider how the quality of men's networks is evaluated. Gender perspectives are led by feminist theories and it is inappropriate to use a 'feminist ruler' (Cancian, 1987) to measure the quantity and quality of men's social networks. As the work undertaken by Davidson *et al.* (2003) identifies, we need to measure intimacy and friendship patterns in the lives of older men in a different way. It needs to be considered that older men may have a wish to have a smaller, closer network of friends and acquaintances and no wish to nurture and support each other (*ibid.*).

An issue for both the UK and Germany is how social policies accommodate the needs of a gradually increasing number of old men who live alone (OPCS, 2002), might experience difficulties in establishing an 'off stage' identity in old age, and might not have maintained good relationships with their sisters or daughters throughout their lives. Men living alone, in same-sex relationships, living apart together, cohabiting, divorced, widowed or married and their resulting social networks form part of the population of pensioners, and the accumulation and maintenance of their social capital can be seen to be key to their well-being in the later stages of life.

Discussion

For the individual, it is the experience of the ageing process that is of greater importance than the chronological onset of old age (Bytheway, 1997). Neither Marjorie nor Renate felt 'old' when they became pensioners and were classified as 'old' by social policy makers in the UK and Germany. They have enjoyed their retirement and as they approached their 80th birthdays they were ready to acknowledge the onset of physical frailty and were learning to adapt and pace their lives. This is an indicator of the onset of personally experienced old age in contrast to a classification determined by social policy. Their personal networks are key to the successful adaptations they have made and may yet experience in old age. As exemplars they do not indicate great differences between the UK and Germany. In fact either could actually be living in the UK or Germany. However, they do provide food for thought relating to the accumulation of social capital and the changing characteristics of the population of old people in both countries, which suggests that European Union policies may be leading to similar outcomes in the UK and Germany.

Marjorie and Renate both formally became pensioners a few years before they moved away from the cities where they had spent most of their working lives. However, it was the physical act of moving to a different geographical location that marked their transition into retirement. Whilst social policy determined their classification as 'pensioners' their actual transition into old age was individually instigated and the relocation can be identified as the marker of this event in their lives. The choice of location was strongly influenced by the distribution of the family network and was related to how they chose to manage and budget their social capital.

Their socio-economic status and the differing 'housing cultures' in the UK and Germany influenced their type of accommodation. Marjorie and her husband live in a bungalow that they own. Accommodation can be identified as a status symbol and potentially also influences the neighbourhood that Marjorie chose to live in. The ownership of the property also provides a financial resource if needed. In the UK, this is not exceptional and does for example mean that for as long as they are physically and financially able to jointly look after house and garden, Marjorie and her husband can maintain autonomy. Here the gender distribution of everyday tasks determines when the support, assistance and guidance of others are required to maintain house and garden. Marjorie's concern is related to her physical ability to do the food shopping. This has been given thought, a contingency plan is in place and she is learning to use the computer in order to undertake her shopping on the internet for delivery to her doorstep.

Were Marjorie and her husband to become more housebound, communication systems would become even more important to maintain contact with their family and social network. Being able to talk to their children and email their grandchildren will enhance their everyday life. Future technological developments in this area will be an important aspect of their lives and their awareness of such developments is likely to ease their acceptance of equipment that might need to be introduced into their households to enable them to 'stay put'. Their cultural capital and the educational aspect of their social capital have enabled them to incorporate modern technology into their everyday life, and in future it will have a key role in maintaining both their social capital and autonomy.

Once assistance is required, the social network that Marjorie and her husband have established is likely to change in character and become a 'support network', which will include formal assistance in the form of paid household help and a gardener. As they have the resources, this could be funded by their personal income and they would probably not expect support from the welfare state. However, as dependency increases there may well be a call on health and social care services to enable Marjorie and her husband to remain in the bungalow. Here, as described elsewhere in this book, the support they receive for health care will be needs-tested and their social care requirements will be means-tested. As a couple they will continue to provide each other with support which may well include taking on the role of carer for the other partner (Arber *et al.*, 1988, Arber and Gilbert, 1989; Calasanti, 2003). It is when one partner is left on their own that the situation changes and when the importance of the individual's social capital increases. If Marjorie were to be left on her own the presence of her son and daughter-in-law living nearby means that family support is available. However, as described by Finch and Mason (1993), negotiating family responsibilities may be complex and it cannot be taken for granted that caring will include the physical 'hands on' help that might be needed. The network surrounding Marjorie may well take on a more formalised structure as services provided by the welfare state are introduced into her life. Marjorie's financial status can be expected to change considerably (see this volume, Chapter 2) and may consequently impact on her ability to maintain her social network consisting of family and friends. It can also be anticipated that the time may come when social capital needs to be cashed in exchange for assistance. This will postpone the time when the social network is extended to include people paid to enter her personal sphere and provide assistance with everyday activities.

In Germany, house ownership is not the norm and many people irrespective of their age live in rented accommodation. On the one hand this means that Renate does not have to be concerned about the maintenance of the fabric of the house or the garden. It does mean, though, that she has to pay rent out of her monthly income. Assistance in maintaining her household to the standard to which she aspires is, not unsurprisingly, paid for from her income. As a divorced woman living on her own, Renate does not have the mutual support Marjorie and her husband experience. This mutual support has both emotional and financial aspects as well as bringing assistance in everyday activities. Renate is accustomed to being on her own and does not have the intimate emotional support experienced within couple relationships. Her financial situation is related to her divorce settlement and the career she managed to develop following the divorce. This has provided her with an adequate income since reaching retirement age. This cannot always be seen as the norm for divorced women in either country. It was Renate's personal drive and ability to establish a career that enabled her to secure a comfortable income as a pensioner. Had she not achieved this she would depend on support from social services to complement her pension (this volume, Chapter 2).

To date, the issue of divorce within the population of old people has not been widely discussed. As Arber *et al.* (2003) point out, this raises an important gender issue. In the UK both divorced older men and women are financially more disadvantaged than their widowed counterparts. However, divorced women are at greater risk of experiencing poverty than divorced men. Amongst men living alone it is divorced men who represent the most disadvantaged group. The financial resources of an older person will invariably also impact on their ability to maintain a social network. If, for example, Marjorie or Renate were unable to afford to reciprocate invitations to meals or to participate in joint outings, their social networks could be expected to be much smaller. Here marital and financial status are closely linked and can be seen to impact on an individual's ability to maintain a social network.

Assistance in everyday living is potentially more problematic and it can be expected that Renate might need outside assistance where Marjorie or her husband do not because they can help each other and their son lives nearby. Renate may need to adapt her lifestyle more radically and, for example, has chosen to wear clothing, eat meals or undertake activities that she feels competent to deal with independently in order to maintain her autonomy. However, her close network of friends and relatives means that she is not left isolated. Emotional support from her children, nephew and nieces is always available over the phone. In addition her brother, sister-in-law and friends are physically close by and available in times of need, and within her network of family and friends, this is reciprocal. Because Renate lives on her own and is part of a network of women who live on their own, a greater awareness of when help and support is required can be expected. Within the group difficulties that one or the other is experiencing may be discussed and solutions jointly found. Here women's mutual nurturing could take place in a way that men may well reject. Renate exemplifiesthe fact that in old age, a social network based on friendship is important to old women's well-being. The global distribution of families exemplified by both Renate and Marjorie indicates that in future, home and hearth may no longer be the sole centre of old women's lives.

Should Renate have a problem with her shopping then the network of family and friends would find a solution. In this instance technology, in the form of internet shopping, would probably fail because it is not part of Renate's present life. It would be a major hurdle for her to overcome should she be confronted with it. Potentially it could also be the network of family and friends who find a formal carer who would be paid for, if a claim for long-term-care funding at home (see this volume, Chapter 9) were to be made. Only if Renate were unable to fund her care needs from her own income plus the finance available from her long-term care insurance would she approach social services (*ibid.*). Means testing would take place and should financial support be made available from social services, her children would also undergo a means test and be called upon to pay money to social services. This is in contrast to the UK where Marjorie would be evaluated as an individual and not as the member of a family. This is indicative of the family-orientated focus of German social policies. For both women, the need for assistance in activities of daily living would potentially alter the dynamics and consistency of their network as formal care givers became part of their everyday life and consequently their social network.

At present women account for the largest proportion of people who have attained a very advanced age and are living alone, and there is a feminisation of the oldest old who live alone (Tews, 1990; Naegele, 1991; Clemens, 1993). Single women in need of care are severely affected by issues related to the provision of care in old age. In this instance, the fate of women living alone in the UK and Germany is similar and they can expect to experience institutional care. However, the next cohort of the oldest old is starting to present a different picture, and men are increasingly having to deal with similar issues in both countries. As the life course of men and women becomes more egalitarian, so do their experiences in old age. This would mean that in future men will also see themselves confronted with formerly 'typical female' problems of old age.

In future, partnerships in old age can be expected to become even more of a key feature in the characteristics of pensioners than has been the case to date. This could be seen to defuse the gender issue. Especially for married pensioners, the division of labour established throughout the marriage and working life appears to be renegotiated. For Marjorie and her husband, the redistribution of household duties continues to be gender-specific: her husband's responsibilities include shopping and the maintenance of the house whist she does the cooking and house-work. Gardening is her hobby but he cuts the lawn and maintains the pond. Whilst exploring the social world of men, Davidson *et al.* (2003) identify how men take over cooking meals and justify this role by point-ing out that some of the best chefs in the world are men, indicating that household chores are done by the person willing and able to do them best, irrespective of gender. Marjorie and Renate's children are part of the 'baby boomer' generation and will soon join their parents and become pensioners. As this new generation become pensioners, changes in family life and the division of labour can be expected to become more obvious.

Social networks for old men in both the UK and Germany appear to be eclectic. As these networks can be expected to reduce with increasing old age, men living alone are at greater risk of experiencing isolation. Whilst Renate, living on her own, has a supportive network of friends and family, this may not always be the case for a man living alone.

Gradually gender equality and conformity in European policies is diminishing the differences between genders in the UK and Germany. For old people the main difference both by gender and by country is related to financial resources and the resulting impact these have on an individual older person's quality of life. The next cohort of young old people are the 'baby boomers' who were not involved in World War II but actively campaigned for gender equality in the 1960s. It can be expected that gender differences, influenced by government policy, will consequently have influenced the life course of men and women. The resulting outcome for the lives of those born after World War II remains to be seen. Potential policy changes relating to the onset of life as a pensioner are already taking place in both countries and in future men and women born in the same year will reach old age together and at a chronologically later stage in life. The outcome of health care policies over the last fifty or more years will potentially also postpone the onset of physical frailty that, for example, Marjorie and Renate are experiencing as octogenarians.

Policies that impact on the health and wealth of old people in the UK and Germany will support and enable the accumulation of social capital through the creation and maintenance of networks. This can be seen to be a valuable resource that will add to the well-being and quality of life experienced by old women and men in both countries. However, there is a danger that in both countries, old single men who never married, are divorced or widowed, may lack social capital and be overlooked. This is in contrast to old single women who may have a restricted income but are more likely to have a well functioning social network and therefore have sound resources of social capital.

Acknowledgements

We would like to express our gratitude to Sara Arber for her comments during the writing of this chapter, and especially to thank 'Marjorie' and 'Renate' (they know who they are) for permitting the use of their biographies in this piece of work.

6
Minority Ethnic Elders in Germany and the UK

Vera Gerling

Introduction

In many European countries the development and delivery of adequate personal social services for minority ethnic elders have become important socio-political issues. This is due to the fact that in the context of migration triggered by colonial connections, poverty, insecurity and job search, the European countries of today are increasingly being influenced by population groups who have immigrated particularly since the 1950s (Filtzinger, 1995). Even though the minority ethnic elder population is relatively small today, it will increase rapidly within the next decades. Apart from general questions concerning the form and the extent of social, cultural and political integration of minority groups within the societies in which they have settled, community care systems are challenged by their very presence. Care providers face the task of opening up their services for minority ethnic elders with their different cultural, ethnic, religious and linguistic backgrounds. Germany and Great Britain are characterised by clear differences in their migration history, but the two countries face comparable challenges in demographic and socio-cultural development.

Numbers, ethnic background and life circumstances of minority ethnic elders

The statistical classifications of immigrant population groups vary significantly between the two countries, which makes comparisons difficult. In Germany, the share of the so-called foreign population (defined as people who do not have German citizenship) has stayed the same over the period 2000–05 (about 7.3 million people), representing some 9 per cent of the total population in 2003. The largest foreign groups were people from Turkey (about 1.8 million), Italy (about 548,000), Serbia and Montenegro (about 507,000), Greece (about 316,000), Poland (about 292,000), Croatia (229,000) and the Russian Federation (179,000) (Statistisches Bundesamt, 2004a).

According to the UK Census 2001, the minority ethnic population was 4.6 million, representing a 7.9 per cent share of the total population (Office for National Statistics, 2003). The largest minority ethnic groups were Indian (about 1.05 million), Pakistani (about 747,000), those of Mixed ethnic backgrounds (about 677,000), Black Caribbeans (566,000), Black Africans (48,500) and Bangladeshis (283,000).

In comparison with 1991, the minority ethnic population in Great Britain had grown by 53 per cent from 3.0 million to 4.6 million, which was partly a result of the addition of the category 'Mixed ethnic groups' (Office for National Statistics, 2003).

Germany

In Germany the proportion of foreign elders was relatively small as of 2003 (about 10 per cent share) due to the comparatively young age structure of the population. At the end of 2003 there were about 760,000 foreign elders in the age group 60 years and above. Comparing the foreign elder population with that in 1995 (467,000), the figure had increased by 70 per cent.

Projections estimate an increase in the number of foreign senior citizens to about 1.1 million in 2010 and to almost 2.1 million foreign elders aged 60 years and above in 2020 (Schopf and Naegele, 2005). Thus, foreign senior citizens are expected to be the most rapidly increasing population group in Germany (Bundesregierung, 1993). In addition to the foreign elders registered in statistics, there are also about 420,000 (the figure relates to 1999) so-called resettled senior citizens (aged 60 years and more) from the Eastern parts of Europe and Russia. Statistically and policywise, they are treated as Germans but they too have experienced migration and different cultural socialisations (Dronia, 2000).

The largest groups of foreign elders aged 55+ years come from Turkey (289,000), followed by Italy (105,000), Serbia and Montenegro (89,000), Greece (72,000), and Croatia (69,000) (Schopf and Naegele, 2005).

Great Britain

In Great Britain, the largest numbers of elders from black and ethnic minorities originate from Ireland and the New Commonwealth (Caribbean islands, India, Pakistan, Bangladesh) (Coleman, 1996).

The different statistics which aim at categorising minority ethnic groups concentrate on criteria such as skin colour and national/ethnic origin. Because, until 1962, most immigrants from the Commonwealth automatically gained British citizenship when entering the country, they are called ethnic minorities. To distinguish them from the predominant white indigenous population, they are also subsumed under the term 'blacks'.

In the age group 65 years and above, there were about 556,000 elders from black and minority ethnic groups living in Great Britain in 2001. They were comprised of the following groups: White Irish (171,559), Other White (148,811), Mixed (19,971), Indian (69,374), Pakistani (31,037), Bangladeshi (9,131), Other Asian (12,778), Black Caribbean (59,987), Black African (11,226), Other Black (3,161), Chinese (12,416), Any Other Ethnic Group (6,600). The group of all non-white elders comprised about 236,000 people (Office for National Statistics, 2003; ONS / General Register Office for Scotland, 2000).

The number of people in minority ethnic communities who are aged over 60 will increase to nearly 1.8 million by 2016 (Age Concern England, 2001b). In general, the numbers are estimated to be increasing with a predominance of elders from India and the Caribbean islands (Patel, 1994).

Elders from black and ethnic minorities in Great Britain were comprised of the following groups: Indian (about 57,400), Black-Caribbean (about 54,300), Pakistani (about 17,600), Other-other (about 14,600), Chinese (about 9,000), Other-Asian (about 8,000), Black-African (about 5,700), Bangladeshi (about 5,300) and Black-other (about 3,700) (Age Concern England and Commission for Racial Equality, 1995).

In contrast to the majority ethnic population, there are more male minority ethnic elders in both countries, although clear distinctions exist between different ethnic/national groups. In Germany, elders from Morocco (83 per cent) and from Tunisia (80 per cent), and in Great Britain elders from South Asia, have the highest shares of men (Owen, 1993; DZA *et al.*, 1998).

The regional distribution of elders from black and ethnic minorities in Germany and Great Britain is quite similar. Due to work migration, the highest concentrations are in the northern parts of old industrial conurbations. On the local level, the compositions of minority ethnic groups differ significantly in both countries (Owen, 1993, 1996; Deutscher Bundestag, 1998).

Life circumstances of elders from black and ethnic minorities

Elders from minority ethnic groups face greater disadvantage than those from the majority elder population in both countries (Krieger, 1993; Naegele, 1993). Disadvantage arises in income maintenance (Age Concern and Commission for Racial Equality, 1995; DZA *et al.*, 1998; Freie und Hansestadt Hamburg *et al.*, 1998), in housing (Atkin and Rollings, 1994; Age Concern and Commission for Racial Equality, 1995), in leisure, regeneration and health (British Medical Association *et al.*, 1995; Richards and Abas, 1996; Bhatnagar and Frank, 1997; Han, 2000; Bundesministerium für Familie, Senioren, Frauen und Jugend (BMFSFJ) and Matthäi, 2004) and in provision and participation. However, regarding citizenship and legal rights, the UK's minority ethnic elders are better off than their peers in Germany (Baringhorst, 1993; Gerling, 2001).

Minority ethnic elders also face greater isolation, and reduced educational and learning opportunities compared with their majority-group elder peers (Dietzel-Papakyriakou, 1993). Family and/or community support is variable for minority ethnic elders: the notion that 'minority elders are looked after by their families' need not hold good in each and every family: after all, minority families are integrating and the idea of care within the family has been rapidly changing with majority families for some time (Gerling, 2001).

With a rising minority ethnic elder population facing general sociocultural and structural changes within their groups compounded by experience of disadvantage, it is to be expected that there will be a greater need for support and care services in both countries.

Care needs of minority ethnic elders result from two processes: first, universal biological and physical ageing processes no different to those of majority senior citizens. Second, migration, experience of discrimination and disadvantage, together with differences in culture, language and faith give rise to needs that are different.

Legal frameworks and their impact on the provision of personal social services for minority ethnic elders

The relevant legal frameworks of services for minority ethnic as well as for majority elders explain the prevailing system of community care for elders and its legal basis. For Great Britain, the anti-discrimination legislation is of crucial relevance.

Community care for elders in both countries aims at securing an independent and self-determined life in old age, which takes priority over institutional care. There are however some differences: in Germany, due to the principle of subsidiarity, voluntary organisations have a limited priority over public providers, except for services in the field of responsibility of care insurance (Bäcker *et al.*, 2000). Although legislation in Great Britain aims at delegating the provision of services to the private and voluntary sectors, the public sector, in contrast to Germany, is still the largest provider of personal social services for elders' health and social care. In terms of providing social services to minority ethnic elders, the minority ethnic voluntary sector plays a predominant role. Due to a lack of public and mainstream voluntary services, minority voluntary organisations have had to step in.

In contrast to Germany where progress on meeting EU directives on 'race equality' has been slow, Great Britain has had a long established history and practice of anti-discrimination legislation since 1965. The Race Relations Act 1976 forbids direct and indirect discrimination on grounds of skin colour, race, ethnic or national origin (Commission for Racial Equality, 1997). Section 71 of the Race Relations Act 1976 has also focused on local authorities (Randall, 1991). The Race Relations (Amendment) Act 2000 expanded the area of responsibility to all public institutions carrying out public functions and similarly to private institutions that carry out public functions such as education. Moreover, in the case of discrimination, the Act also aims at tightening up sanctions (Commission for Racial Equality, 2000).

This legal framework has had an impact on the development of social services for elders from black and minority ethnic groups. Ever since the agenda of social services for minority ethnic elders has gained attention, partly due to the Act and partly due to organisations and elders expressing greater dissatisfaction, additional guidelines have been developed that support public providers of social care in considering the specific needs of elders from black and minority ethnic groups (for example by the then Commission for Racial Equality and the Social Services Inspectorate). Such processes allow for a closer cooperation between relevant actors in the public service sector and minority ethnic groups. The British multicultural approach of anti-discrimination legislation also impacts on the delivery of social services for minority ethnic elders: in contrast to Germany, specific and separate services for elders from black and ethnic minorities are more widespread and not regarded as an obstacle to integration (though government thinking as of 2005 was mixed on this). Following this approach, translated information material as well as translation and interpretation services are also more common than in Germany. From the German perspective, minority ethnic housing associations offering sheltered housing catering to the needs of minority ethnic elders represent a striking difference between the two countries.

British anti-discrimination legislation also influenced the NHS and Community Care Act 1990 which is the legal basis of community care for elders. Following this Act, local authorities must consult certain population groups such as users and carers when developing community care plans (Means and Smith, 1998). In this Act, people from black and minority ethnic groups are explicitly regarded as having specific needs and demands that have to be met. Social services departments in their role of public providers of social care for elders are made responsible for identifying specific needs of minority ethnic elders and for considering different ideas about community care. It is not clear how far the Act has contributed to a better provision of services for minority ethnic elders in general.

Transposing EU Race Equality Directives (2000/43/EC of the Council from 29.6.2000, Article 3c) into national law was a long and slow road for Germany and some other EU Member States. Following specific actions by the European Commission as allowed by the Race Equality Directives, on 29 June 2006 the German Federal Parliament passed the Act of Equal Treatment (Allgemeines Gleichbehandlungsgesetz (AGG)). The AGG became operational following amendments in October 2006. This action resulted in the implementation of the EU Race and Ethnic Origin Directive 2000 and the Framework Employment Equality Directive 2000, although their full implementation is yet to be examined. What is clear is that the establishment of such anti-discrimination measures for equal treatment of all citizens offers Germany a potential to develop its services to its growing ethnic elder population.

The legal basis governing the concerns of foreign citizens in Germany is an immigration law that passed the German Bundestag in the Summer of 2004 and was implemented in 2005. The 'Immigration Act' consists of comprehensive regulations concerning abode, occupational activities, migration of EU citizens, asylum, citizenship and integration of foreigners in Germany.

Compared to former legislation, the Immigration Act is generally more oriented towards the integration of immigrated and immigrating persons. This is mainly attempted by making so-called language and orientation classes mandatory for most foreigners. In turn these classes are oriented to the abilities of immigrants and comprise 630 instruction units, mostly language classes.

Compared to Great Britain, the development of social services for immigrant elders cannot be traced back to legal pressure, although the German Association (Deutscher Verein) explicitly points out that community care for elders has to focus on *all* senior citizens living in Germany. The mandate of community care for elders and its area of responsibility aims at all old people in Germany, i.e. also at immigrant elders (Deutscher Verein, 1998).

Since the 1990s, the public sponsorship of model projects has been a characteristic feature of the German attempt to gather information about the life situation of immigrant elders and about strategies for opening up services. Three projects have contributed to obtaining detailed information: 'Developing Concepts and Strategies for Community Care for Immigrant Elders', 'Adentro! Spanish Speaking Senior Citizens Get Involved', and 'Germans and Foreigners Together: Active in Old Age'.

Recent policy issues, initiatives and projects in Germany, the UK and the EU

Efforts to develop and open up social services for elders from black and minority ethnic groups pre-date legislative imperatives in both countries as witnessed by many centres which serve minority ethnic elders be they in Berlin or London. However these services are uneven with many service gaps. For example, there has been consistent research evidence for both countries that minority ethnic elders are still not well-informed about the existence of public and voluntary personal social services for elders and that the takeup of services is relatively low as between different ethnic groups (Bhalla and Blakemore, 1981; BMA *et al.*, 1995; Freie und Hansestadt Hamburg *et al.*, 1998).

The reason for the low takeup of social services is the existence of access barriers. These include a lack of attention to special ethnic, linguistic, religious and nutritious needs of black elders, a lack of information about their life situations, and a lack of possibilities for reaching them. British scientific literature additionally points out institutional racism, general discrimination and non-attention to the needs of elders in general. On the side of minority ethnic elders, the largest access barrier concerns insufficient knowledge about services and institutions, wrong perceptions about the content and structure of services, certain cultural and religious concepts, language difficulties and poor service experiences (Blakemore and Boneham, 1994; Age Concern England and Commission for Racial Equality, 1995; Askham *et al.*, 1995; Butt and Mirza, 1996; Commission for Racial Equality, 1997; DZA *et al.*, 1998; Social Services Inspectorate, 1998; Gerling, 1999).

On the basis of model projects and general experiences, both countries have developed a list of demands that focus, for instance, on issues such as strategy and policy, planning and service development, commissioning and providing services, care management, inspection and quality assurance, employment and training (Bundesregierung, 1993, 1997; Bundesarbeitsgemeinschaft der Freien Wohlfahrtspflege, 1995; Commission for Racial Equality, 1997; Naegele *et al.*, 1997; Deutscher Verein, 1998; Social Services Inspectorate, 1998).

When we compare local practices in Germany and the UK in developing and providing social services for immigrant elders, we find that for example in the twin cities of Dortmund and Leeds respectively, both cities use similar strategies but that the extent and the level of development is higher in Leeds (Gerling, 1999).

Recent policy initiatives and projects in Germany

Compared to the 1990s, the issue of age and ethnicity has gained more weight at the political, academic and practical levels in the new millenium. There has been an increase of projects at the local level, better cooperation of different actors and a stronger support by the German government.

In 2000, associations and institutions working in the fields of 'care for elders' and 'migration/integration work' networked and developed the so called 'Campaign for a Culture-Sensitive Care for Elders'. Together, they wrote a 'Memorandum for a Culture-Sensitive Care for Elders' and developed a 'Checklist for a Culture-Sensitive Care for Elders'. These documents were presented and discussed publicly in June 2002. The campaign aimed at sensitising relevant actors and the general public to the need to adapt elder care institutions and services to the needs of minority ethnic elders to enable them to live a life in dignity. A further objective is to develop and provide appropriate frameworks. The official campaign was launched on 1 October 2004 in Berlin. It offers information packages for interested actors and has its own road show.

In 2002, the Federal Ministry for Families, Seniors, Women and Youth (BMFSFJ) established a clearing, contact and information office

(Informations- und Kontaktstelle Migration (IKOM)). This initiative concentrates on age and ethnicity and collects information on projects, initiatives, and literature that can be used by any interested person. The same ministry also provided funding for a research project that focused on the group of older migrant single women, an area that has been neglected in practical projects and research. Based on qualitative interviews it became clear that this group itself is very heterogeneous but that social circumstances and integration largely depend on educational opportunities and cultural resources. The social integration of older migrant single women is facilitated by education, social cohesion, orientation towards advancement and German language abilities (BMA and Matthäi, 2004).

At the beginning of 2004, there were some 30 different projects, characterised by different core activities, operating at the local level in Germany. They provide the following services and support:

- intercultural socialising and exchange of experiences;
- advice and counselling for minority ethnic elders;
- opening up and adapting mainstream services of community care to the needs of minority ethnic elders;
- biographical approaches;
- self-help, campaigning and social participation of minority ethnic elders;
- development of adequate information material.

These projects differ in terms of their content, target groups, resources, institutional setting, form of organisation, and methodology.

With regard to the content, there has been greater emphasis on care activities such as dementia services or sheltered housing.

Concerning target groups, some of the projects concentrate on minority ethnic elders themselves and others on staff of outpatient services, institutional care or advice services. Only a few projects concentrate on both target groups. There are some intercultural approaches that focus on majority and minority ethnic elders. The target group of minority ethnic elders can be differentiated by nationality or cultural background and by gender. Most projects aim at minority ethnic elders from former recruitment countries, some focus on resettled elders from Eastern Europe and the former Soviet Union.

With regard to the institutional setting, most projects are still linked to one of the German welfare associations or to the local public sector.

Only a few projects are offered by self-help groups or organisations of minority ethnic groups.

One project that was funded by the Ministry of Health, Social Affairs, Women and the Family of North Rhine-Westphalia illustrates the development and use of information to increase access to services. The project was supported and strengthened by extensive public relations work, run by the district of Unna (near Dortmund) and a minority ethnic organisation.

Elders from Turkey were given 12 information units on issues such as German long-term care insurance, health and social care services, and consumer advice. All sessions were conducted in Turkish, some of them in community centres and mosques.

At the same time, staff of outpatient services, institutional care or advice services became qualified in the special (care) needs of minority ethnic elders. Eight units focused on topics such as life situation of minority ethnic elders in Germany, culture-specific perceptions of ageing, culture-specific attitudes regarding care, illness and body concepts, access barriers to minority ethnic elders with regard to mainstream services, and an excursion to an institution that offers care for minority ethnic elders. The project was evaluated and its approaches were proven quite successful (Gerling, 2004b).

In summary, it can be stated that research and projects have been broadened in recent years and that some improvements have been made. In general, there is greater sensitivity regarding the needs and demands of minority ethnic elders. However, services are still piecemeal and many service providers experience difficulties in identifying and reaching minority ethnic elders.

Some target groups, especially smaller minority ethnic groups, are still being neglected. There is not much information on the needs and demands of one of the largest groups, namely resettled senior citizens from Eastern Europe and the former Soviet Union (Bundesregierung, 2005).

Recent policy initiatives and projects in the UK

Policy issues

Several major developments have shaped the agenda of care for older people. Consequently these have had an impact on minority ethnic elders. We illustrate only one example of health policy that has guided the work on older people.

The National Service Framework for Older People was published by the Department of Health in 2001. Based on a ten-year programme, 'it is the first ever comprehensive strategy to ensure fair, high quality, integrated health and social care services for older people' (Department of Health, 2001a, p. 1). The National Service Framework addresses the needs of older people and sets out eight standards for their care people across health and social services in the following fields: (1) rooting out age discrimination, (2) person-centred care, (3) intermediate care (4) general hospital care, (5) stroke, (6) falls, (7), mental health, (8) the promotion of health and active life, (9) medicines and (10) local delivery. The National Service Framework acknowledges that 'older people are not a homogeneous group. The proportion of older people from black and minority ethnic communities is small but growing. All services should reflect the diversity of the population which they serve' (Department of Health, 2001a, p. 5). The specific needs of black and minority ethnic elders are especially addressed in the sections on stroke and mental health. The National Service Framework was reviewed by the Healthcare Commission, the Commission for Social Care Inspection, and the Audit Commission. Their report illustrates some developments but also considerable gaps in local authority and health service provision regarding minority ethnic elders' care and quality of life. Old age and mental health has long been regarded as a 'neglected' area.

In 2005, the government published the policy document *Delivering Race Equality in Mental Health Care* (Department of Health, 2005a), of which the lead author was Professor Lord Patel, which includes a five-year action plan for reducing inequalities. The programme will be based on three pillars: (1) more appropriate and responsive services, (2) community engagement that will be supported by 500 new community development workers, and (3) better dissemination of information and improved monitoring of ethnicity. Concerning the needs of black and minority ethnic elders, it is recommended that all mental health services take account of the their language and interpretation needs. Consequently the Department of Health, responsible for implementing the policy and its recommendations, has invested in minority elder mental health-specific innovations to stimulate service responses.

Health issues

Apart from these policy developments, progress has also taken shape in the form of research and service responses.

In 2004, a three-year Inquiry into Mental Health and Well-Being in Later Life was started by the Mental Health Foundation and Age Concern England. The main objectives of the inquiry were (1) the development of a broad and in-depth understanding of mental health and well-being in later life; (2) the improvement of the quality and range of services for older people, their families, carers and friends; (3) to influence national and local policy and planning; (4) to increase general awareness and foster ongoing debate, and (5) to champion older people as experts in their own mental health and well-being. The inquiry also focused on the special needs and views of black and minority ethnic elders (Age Concern England, 2004a). The first step aimed at understanding what helps diverse groups of older people to maintain their mental health and well-being throughout later life. For this, a literature and policy review was undertaken, a 'Call for Evidence' was launched and over 5,000 questionnaires were sent out (Age Concern England, 2004c). Since there were almost no responses from black and minority ethnic elders, three specific consultation sessions with members of this group were set up (Age Concern England, 2005a).

In 2004, the Parkinson's Disease Society organised a one-day workshop which was the starting point for the development of a practical guide for staff working locally with Black and Asian groups. The guide tries to support attempts to reach out to people from these groups (Age Concern England, 2004b). The charitable organisation CancerBACUP (2004) published a report, *Beyond the Barriers*, that was intended to help everyone involved in cancer care to deliver high-quality cancer information to black and minority ethnic communities. The report reflects the experiences of people working in over 30 projects in the field of cancer and with these communities (Age Concern England, 2004b). Apart from the actions described above, a number of other workshops and conferences on health and social issues of black and minority ethnic elders took place throughout the UK (organised for instance by the Policy Research Institute on Ageing and Ethnicity (PRIAE) and the King's Fund).

Recent initiatives of key organisations for older people

In 1998 PRIAE, a specialist independent institute, was set up to develop and improve policy, research, information and services concerning minority ethnic elders.

In 1999, Age Concern England established an Ethnic Minorities Steering Group during its major initiative 'Millennium Debate of the Age' because the existing five theme groups for this debate, (1) values and attitudes, (2) work and lifestyles, (3) the built environment, (4) health and social care, and (5) paying for age had, with the exception of group (1), not considered the issues of black and minority ethnic elders. On the basis of four half-day meeting and brief papers by members, a report was written that gave an overview of research findings on black and minority ethnic elders in general and on findings related to the five themes mentioned above in particular. It ended by giving theme-specific recommendations for further work (Age Concern England and Ethnic Minorities Steering Group, 1999).

The National Black and Minority Ethnic Elders Forum was established in 2002 following an Age Concern conference. It aims to ensure that the voices of older people from black and minority ethnic communities are heard; encourage and acknowledge the contribution that black and minority ethnic elders make to society; and, seek ways of ensuring that their needs are met (Age Concern England, 2003a). The Forum publishes a newsletter that informs readers on relevant issues of age and ethnicity. Together with Age Concern's training arm, it also offers free training courses for black and minority ethnic elders with the aim of improving their ability to express their opinions and to contribute to consultations and user involvement opportunities (Age Concern England, 2003b).

Help the Aged set up a three-year Minority Ethnic Elders Falls Prevention (MEEFP) programme in 2004, funded by the Department of Health. The programme aims to raise awareness about the risk of falls among black and minority ethnic elders and those who care for them. With regard to the difficulties of reaching these elders with mainstream fall prevention services, it was decided to involve them and their communities in the design and delivery of special projects. The strategies used in the project relate to: (1) improving health and well-being in later life among the Asian community in Barnet, London; (2) 'taking positive steps' with Cypriot women in Camden, London; (3) a falls awareness road show for Manchester's Chinese community; (4) the making of a Cantonese falls awareness video in Portsmouth, and (5) working with South Asian elders in North London. The evaluation report establishes that over 500 black and minority ethnic elders have been directly involved in the programme (Help the Aged, 2005a).

The issues of minority ethnic elders have also been addressed within the Better Government for Older People (BGOP) initiative, which was set up in 1998 with the support of central government and consists in a partnership of Age Concern, Help the Aged, Anchor Trust,

Carnegie Third Age Programme and Warwick University. Its objectives were to:

- make a difference to the lives of older people as citizens;
- engage older people in decision making;
- cooperate to share experience and knowledge.

The programme was centred around 28 locally based two-year pilot projects in which over 300 organisations were involved. The projects aimed to test and develop innovative strategies to involve older people within a democratic model (Maltby and Rohleder, 2005).

Local black and minority ethnic elders' forums

Many local forums exist, like the Leeds Minority Elders initiative or the East Midlands Black and Minority Ethnic Elders Project. Essentially their purpose is to bring together a range of minority and majority old-age organisations; to raise awareness of mainstream institutions, service providers, funders and policy makers about the needs of black and minority ethnic elders and their organisations, and to offer minority ethnic elders opportunities to participate and inform change. The overall aim is to achieve positive changes in the lives of these elders.

Projects and new publications

Besides these initiatives and programmes, many projects that aim to deliver services for black and minority ethnic elders in different fields have been set up or further developed. Some examples include:

- the Dutch Pot Lunch and Social Club for African-Caribbean Elders in Westminster;
- the Chinese Community Centre for Chinese elders in Soho;
- the publication of free IT learning materials for older people who read Urdu or Punjabi, in order to make the internet accessible for more Asian elders (Age Concern England and Barclays);
- Dhek-Bhal – services for South Asian elders in Bristol and South Gloucestershire;
- a reminiscence art festival and symposium ('The Place Where I Grew Up') held in Spring 2004 in London;
- the Dudley African-Caribbean Community Care Befriending Service;
- the Pepper Pot Day Centre for Afro-Caribbean elders in London.

Recent publications cover the learning needs of black and minority ethnic elders (National Institute of Adult Continuing Education, 2003), reminiscence work (Age Exchange, 2004), Chinese care home needs in London (PRIAE, 2003), the situation of the Chinese community (including Chinese elders) in Leeds (Law, 2004), perspectives on ageing and financial planning for old age (Katbamna, 2005), housing advice (Help the Aged, 2005b), black and minority ethnic women in the UK (Fawcett Society, 2005) and black and minority ethnic elders' views on research findings (Butt and O'Neill, 2004).

Recent initiatives and projects in the European Union

SEEM I and II

The European project Services for Elders from Ethnic Minorities (SEEM) aimed to develop and promote solutions to improve health and social care service delivery for minority ethnic elders through the exchange of good practice between the different partner cities.

The project, funded by the European Commission over the period 2003–05, brought together eleven official partners, with a focus on consulting and involving elders themselves. Twelve consultation meetings with minority ethnic groups at local level and ten city-specific ethnic elders' partner exchanges took place. As a result of the project, a good-practice checklist was produced that aims at helping other interested actors at local level to improve services for ethnic elders. For instance, in Dortmund, Germany, a multicultural working group was established, which aims at improving services for minority ethnic elders. In Leeds, UK, a 'dementia cafe' for black elders was set up. The finale was a launch event in Brussels where elders directly participated in the proceedings. The evaluation showed that it is possible to involve minority ethnic elders as active subjects rather than passive recipients of developments, even on a short-term European-funded project.

Minority Elderly Care in Europe

The research project Minority Elderly Care in Europe (MEC), funded by the European Commission Fifth Framework Research Programme (May 2001 – April 2004) and designed by PRIAE, produced a major reference text 'to provide detailed multi-country information on the circumstances, experiences, and care needs of minority ethnic elderly' (Patel, 2003, p. xi) and to present primary research results regarding health, social care and demographic data drawn from work with 26 ethnic elder groups, nearly a thousand professionals and 300 voluntary age organisations across ten countries: the UK, France, Germany, The Netherlands, Spain, Finland, Hungary, Bosnia-Herzegovina, Croatia, and

Switzerland. In 2004, first research findings on each of the ten countries were launched at the European Parliament. The MEC project addresses service issues of minority ethnic elders from three perspectives: that of minority ethnic elders themselves, that of health and social care professionals (including managers and planners) and that of the minority ethnic voluntary sector (PRIAE, 2004).

Conclusion

It is clear that the issues of minority ethnic elders are being increasingly addressed in both the UK and Germany. Recently, within the framework of the European Union Project SEEM II, a checklist was developed by partners, which aims to help other actors to set up services for minority ethnic elders at local level (Gerling, 2005). In general, the following issues should be developed further both in the UK and Germany.

Better understanding of the needs of minority ethnic elders

Although a number of research studies already exist at national country level, more research is needed at local level to indicate exact population figures and the structure of community groups. Specific research needs to be carried out on the life circumstances and needs of single groups such as older people from the gypsy and traveller communities in the UK or on so-called 'late emigrants' in Germany.

Policy development

Both in the UK and Germany, the service projects that have been briefly mentioned (and there are many others not mentioned) have one thing in common: they need to be recognised and supported so that they can do better in service delivery. Greater funding is needed and local authorities encouraged to start developing specific services for elders from minority ethnic communities. It would also be beneficial if the relevant national ministries published guidelines on how to develop services for minority ethnic elders and took a stronger role in promoting successful project initiatives at the national level.

Consulting and involving minority ethnic elders and their organisations

Especially for local authorities, there is an urgent need to develop a better relationship with minority ethnic communities and their organisations at local and national levels. However, the preconditions of consultation and user involvement differ greatly between the UK and Germany, the UK having a stronger tradition and experience than Germany, but from which both countries can benefit, as the SEEM project shows.

Service development

There are several ways for an organisation to help minority ethnic elders to overcome barriers and gain access to services. These include a more patient/user-centred approach, provision of information in the user's own language, provision of interpreters, and training for health professionals in cultural, age and race equality. Further resources are needed to ensure that minority ethnic elders have the same chances to live a healthy and fulfilling life. Successful examples and case studies need to be published by the relevant national ministries to ensure that other local authorities can develop similar projects.

7
The Changing Generational Contract Within and Outside the Family: Britain and Germany Compared

Monika Reichert and Judith Phillips

Introduction

For the last ten years there has been a discussion in the media and in social policy whether or not there will be a 'war of generations' or an 'intergenerational conflict'. One important question which arises in this context is whether or not families will be willing or able to support and care for older family members (*Der Spiegel*, 2005). Another question refers to whether or not older people are an economic burden on society (Schmähl, 2002; Motel-Klingebiel and Tesch-Römer, 2004). The answers to these questions from researchers in the fields of social gerontology and/or social policy are clear. For example, in 2000, Attias-Donfut and Arber concluded in a book *The Myth of Generational Conflict* that at the turn of the century, European societies did not show signs of generational conflict (see also Shanas, 1979). An equivalent statement is made by Baltes (2005) who points out that there is a psychological barrier which prevents a war of generations. When it comes to family solidarity and care and support for older people, most researchers and policy makers are also – a least for the near future – quite optimistic (Kohli and Künemund, 2003).

But what are the reasons for a fear of intergenerational conflict and the collapse of family solidarity? What kind of developments can we expect for the future? What can social policy do to strengthen family solidarity and avoid a 'war of generations'? Is a welfare state able to support a growing older population in a way that is equitable to all generations and ages?

The aim of the chapter is to find answers to these questions by mapping out and comparing the generational contract within and outside the family in Britain and in Germany. To do this we firstly outline important socio-demographic trends in both countries. Secondly we examine the implications of these trends for solidarity within the family (micro level) in terms of economic, social and cultural exchanges. In a third step we describe the macro generational contract and then turn to the consequences socio-demographic development will have for the relationship between the generations. Fourthly, in bringing these two levels together the issues are looked at through a policy lens by trying to outline the current policy debate. Fifthly, we formulate the challenges that British and German societies have to face in the future at the micro and the macro levels. The last and sixth part of the chapter suggests lessons for each country, giving some examples of good practice in how to support family and intergenerational solidarity.

First of all, two important definitions. There are a number of definitions and meanings of 'generation'. It has been used to distinguish cohorts (people born in the same year or group of years), lineage (parent, grandparent, child), the length in years between parent and child, or 'historical time' where people lived through the same 'historical' periods such as the Second World War. For an examination of intergenerational relationships on a micro level we refer to the definition of 'generation' as lineage; and on a macro level we refer to the definition of 'generation' as members of different birth cohorts.

Background data: important socio-demographic trends

Due to declining birth rates and a longer life expectancy the population in Britain and Germany is ageing. In Britain 20.5 per cent of the population and in Germany 23 per cent are over the age of 60 (European Commission, 2005). The fastest rate of growth is among those over 80 years of age. According to data from the European Commission, 3.6 per cent of the German population is over 80 years of age; in Britain the figure is 4 per cent. Because the risk of becoming dependent and in need of care is especially high in advanced age, this group of older people is more likely to require health and welfare services. To sum up the demographic development, we can say that we observe a so-called 'threefold ageing' ('*dreifaches Altern*'; Tews, 1993) in both countries: rising absolute numbers of older people, their relative proportion in the population as a whole also rising, and a rising number of the very elderly or 'old old'. In addition, within the group of older people we increasingly find a cultural differentiation. Britain and Germany are facing the fact that people with different ethnic backgrounds (in Germany, migrant workers from Turkey or Southern Europe; in Britain, West Indians, Chinese, Afro-Caribbeans and Bangladeshis) are ageing in these countries (Chau and Yu, 2000; Phillipson *et al.*, 2000a; Schopf and Naegele, 2005).

As one result families now have a 'beanpole structure': a move from a pyramid shape, with few living grandparents and many children, to multiple generations of grandparents and great-grandparents, rather than large families of children and siblings within the same generation (Bundesministerium für Familie, Senioren, Frauen und Jugend (BMFSFJ), 2003). An elongated, changing structure means that support and help can go up and down the line but divorce, remarriage and childbearing complicate these generational patterns, as such exchanges are no longer dependent on biological or marriage ties. The changes to the family make any definition fraught with difficulty. Families are not necessarily anymore a group of people who are connected biologically. Rather the family can be seen 'as an interdependent group system that may consist of the biological or adoptive and/or influential others' (Smith, 1995, p. xi). In other words: families are becoming more complex.

Due to the two other mega-trends of 'individualisation' and 'pluralisation' (Beck, 1986) we also observe changes in the behaviour, attitudes and lifestyle of people in Britain and Germany which likewise have consequences for families and for the intergenerational relationship on the macro level. These changes – 'individualisation' refers to independence, privacy and post-materialist values, and 'pluralisation' to different forms of living – are mirrored in a growing number of:

- persons who live alone all their life (lifelong singles);
- never-married and/or childless couples;
- divorces and remarriages;
- family members living at regional distances from each other;
- last but not least, women in the labour force.

Consequently, the issue of who takes care of members across the generations and within the family, given the traditional carer role of women, is of growing importance in both countries. But the trends mentioned above not only have an effect on family relationships. Because a continuation of the demographic development – growing numbers of older people and declining numbers of younger people – is expected for the future, problems for social security systems and for households are foreseeable (see below).

The micro level

Family relationships

In the following we would like to examine the *status quo* with regard to support and help within families but also with regard to non-kin intergenerational contacts. First we turn to family relationships. In Britain and in Germany there is much evidence that supports the fact that in older age care giving and support rest within the family (Szydlik, 2000; Lowenstein and Ogg, 2003; Meyer, 2004; BMFSFJ, 2005b).

A good example of intergenerational family support comes from the OASIS study where Germany and England (not Britain as a whole) were compared in relation to measures of solidarity, conflict and ambivalence in an EU-funded study in 2000 (Lowenstein and Ogg, 2003). In this study adult children were asked whether or not they had provided (and received) any of a number of types of help to (and from) parents during the previous year. As shown in Table 7.1, the adult help rates were high in England and Germany: 87 per cent and 83 per cent

Table 7.1 Help and support to and from parents (75+) by domain and country from the adult child's perspective (%)

Item	England	Germany
Provided help *to* parents		
Emotional support	62	74
Transport/shopping	45	49
House repair/gardening	31	31
Household chores	29	34
Personal care	5	9
Financial support	14	7
At least one type of help	87	83
Received help *from* parents		
Emotional support	39	53
Transport/Shopping	6	0
House repair/gardening	2	3
Household chores	4	4
Child care	n/a	4
Personal care	1	1
Financial support	8	11
At least one type of help	44	54

Note: 'At least one type of help *from* parents' does not include help with child care, as this item was not available for the English respondents.
Source: Lowenstein and Ogg (2003).

respectively of children reported that they had helped/supported parents in at least one of the listed domains during the previous 12 months. Emotional support is clearly the dominant type of support between adult children and their older parents: 62 per cent of adult children in England and 74 per cent in Germany who had at least one parent aged 75 and above had provided emotional support to them. English and German parents do likewise in return – although to a greater extent in Germany than in England (53 per cent vs 39 per cent). However, it should be noted that the children's and parents' perspectives cannot be compared directly, because some parents may get (or give) help from (or to) several children. In addition, the parents included in the OASIS study are relatively old, so their ability to help their children is limited in some domains.

Beside emotional support another country difference refers to financial matters. Financial support flows more from older to younger generations in Germany than in England. In England the pattern is in the opposite direction, possibly in response to lower pension levels and higher poverty rates among elders (see also this volume, Chapter 2). The finding supports the crowding-in hypothesis suggested by Kohli (1999) and Künemund and Rein (1999): generous pensions may enable the older generation to contribute and reciprocate in family exchange, thereby encouraging, rather than discouraging, family ties and contacts.

But what expectations of adult children to provide support for elderly parents (filial obligation norms) do people have in England and Germany? Are cultural and policy differences between these countries reflected in filial norms? Based on a scale developed by Lee *et al.* (1998), support for filial norms is measured in the OASIS study as the percentage of agreements with four statements:

1. Adult children should live close to their parents so that they can help each other.
2. Adult children should be willing to sacrifice some of the things they want for their own children in order to support their ageing parents.
3. Older people should be able to depend on their adult children to help them do the things they need to do.
4. Parents are entitled to some return for the sacrifices they have made.

Two statements refer to what adult children ought to do, and two articulate elderly parents' expectations. The results showed that there were substantial minorities who did *not* accept filial obligation norms,

Table 7.2 Normative solidarity: agreement with filial obligation norms, by item and country (% in agreement*)

Item	England	Germany
1 Should live close	30.7	40.2
2 Should sacrifice	46.6	35.5
3 Able to depend on	41.0	55.2
4 Entitled to returns	47.9	26.1

* Agree or strongly agree.
Source: Daatland and Herlofson (2003).

i.e. they did not agree with any of the four statements. However, the rate of disagreement was higher in Germany than in England (34 per cent vs 25 per cent). A closer look at the actual expression of norms reveals the following results (see Table 7.2): if the option of cohabitation is involved, a higher percentage of German than English adult children support this statement (40.2 per cent vs 30.7 per cent). Country differences also emerge for all the other items. For example, Germans were more affirmative about the proposition that older people should be able to depend upon their children for help (55.2 per cent vs 41.0 per cent). On the other hand, they were less likely than the English to agree that children should provide support even at the expense of the well-being of their own children (35.5 per cent vs 46.6 per cent) and that parents are entitled to some return (47.9 per cent vs 26.1 per cent).

In this context it is important to note that although both countries in essence have traditions of the 'nuclear' family, there is a difference with regard to family legislation. In Britain there is no legal obligation to care between adult generations – the social policy in that country is based on individual needs and rights. In contrast, Germany belongs to a more familistic tradition – here there are formal family obligations to provide support to parents and children. It must be taken into consideration that, from the perspective of potential users, professional care services just represent a 'second choice' and that generally, especially with older persons, there is an inhibition against, or even rejection of, professional care services (Giese and Wiegel, 2000), so that these are used only if there are no other options (Schneekloth and Müller, 2000).

An interesting question refers to the welfare state orientation, i.e., the responsibilities of the welfare state and the family in three domains of social policy and services for older people: financial support, instrumental support and personal care. Surprisingly, the English and the Germans were almost univocal about the role of the welfare state: only a minority saw the welfare state as mainly or totally responsible in all three domains. Although the role of the family was upheld in both countries, the data suggest that the Germans and the English favour equal responsibility as between the welfare state and the family (Daatland and Herlofson, 2003). It should be mentioned that age (or cohort) differences were moderate for both filial obligation norms and welfare state orientation.

When it comes to preferences about the providers of long-term help the OASIS data show that in comparison to other European countries England and Germany take an intermediate position. In this context it has to be kept in mind that subjective norms and care opportunities influence these preferences. In addition, Blinkert and Klie (2000) found that social class is also an important indicator whether or not professional care services are used. For the future it is expected that the primary responsibility for older people will more and more be placed on public services (Herlofson and Daatland, 2001).

Ambivalence and conflict

Norms are relatively easy to capture in surveys of intergenerational relations and solidarity. A more difficult concept to measure is that of ambivalence. Ambivalence can be defined as 'mixed feelings'; for example, a parent feeling neither emotionally close to nor distant from a child. Szydlik (2000) points out that intergenerational relationships can be characterised in principle as both: ambivalence and conflict.

The OASIS study found that ambivalent (and conflict) relationships were reflected during periods of transition in family life, particularly where older people were moving toward dependency and where older parents and adult children were attempting to negotiate roles. Differences between Germany and England were found in relation to relationship styles. English parents reported high rates of 'steady' relationships – those that combine affection with little conflict. Ambivalent relationships were most evident in Germany but England reported higher levels of distant relationships. German parents differed from English in that a majority of parent–child relationships were either ambivalent or distant (see Table 7.3).

What this analysis shows is that inter-country differences in the style of parent–child relationships can be demonstrated, but it is difficult

Table 7.3   Parent–child relationship type by country (%)

Relationship type	England	Germany
Close	27	12
Steady	40	29
Ambivalent	11	41
Distant	23	18
Base	697	708

Source: Lowenstein and Ogg (2003).

to explain different family styles outside of an historical framework. One of the explanations for the apparent generational gap between current cohorts of older parents and their adult children in Germany may have something to do with the polarisation (along generational lines) of traditional/radical attitudes that occurred in the 1960s. However, in reviewing interview material in depth from the OASIS study of dyads of parents and adult children experiencing a transition to dependency, it is clear that Germany has a slightly stronger emphasis on family support, which perhaps reflects the situation that children there have a legal obligation to their parents. Additionally what the data show is that the extended family has maintained cross-generational cohesion despite social change. This confirms findings by Lorenz-Meyer (2004) that many young adult children in Germany perceive prospective parental care as a structurally ambivalent situation; the extent of their ambivalence could only be understood by their position within the family and in societal relations of power, the sources of children's ambivalence stemming from this power differential between them and their parents. This was also reflected in a study in Wales (Maio *et al.*, 2004). In Germany, findings on ambivalence also show the importance of gender and generation (Lüscher and Lettke, 2004).

In relation to conflict, low levels were found in both countries, even during times of transition. However, conflict was an important factor in intergenerational life, with one-third to a half of older parents in the OASIS study sample in Germany and England reporting the existence of some conflict and tension in their relationship with their children. The issue of older people's 'autonomy' was a driver for such conflict although this took on a 'positive' aspect in that children wanted to help their parents but their parents did not want to burden their children.

Factors that influence the extent of family solidarity

The kind and amount of help and support which is exchanged between the generations is influenced by different factors, such as having children and/or family members; geographical distance between family members; living arrangements and household composition; and number of family members who are in need of support and employment. For example, instrumental support especially can be given far more easily when family members are living in the same household or in co-residence. In Britain, in previous studies on mothers and daughters who care (Lewis and Meredith, 1988) and on the family life of older people (Sheldon, 1948; Townsend, 1957; Young and Willmott, 1957), proximity played a crucial part in the provision of care, both emotionally and practically. Households of older people were often complex and often multigenerational. Intergenerational reciprocity of care occurred on a daily basis. Older people in these communities were surrounded by others with whom they had close and supportive relationships, and they lived essentially within what Frankenberg (1966) was later to term 'an environment of kin'. At the same time the Austrian sociologists Rosenmayr and Köckeis (1965) formulated the hypothesis of 'intimacy at a distance'. According to them, intergenerational co-residence reflects economic or occupational requirements, rather than emotional closeness between generations. Compelled by economic necessity, co-residence may be fraught with tensions. Instead, separate households may foster a high level of contact and assistance between generations.

In more recent studies of the same areas in Britain (Phillipson *et al.*, 2000) and in Germany (Kohli *et al.*, 2000), children (as well as other relatives) maintained separate households with generally dispersed families, with only one child close by (within four miles). Regular contact (weekly or more) was maintained, crucially through the car and the telephone, confirming the hypothesis of 'intimacy at a distance'. With the introduction of mobile technology, geographical proximity may no longer be the over-riding factor in contact and support between generations. It can be argued that globalisation in this sense has compressed space and time, yet the provision of hands-on care remains a task constrained by distance.

One factor which can also hinder younger family members from giving support and care to their old relatives is gainful employment. However, studies in Germany and Britain show that more and more (female) care givers try to combine both – employment *and* elder care (for an overview, see Phillips, 1995; Reichert, 2003). For example, in Germany 34 per cent of all carers of employable age, i.e. between 15 and 64 years, are working full or part time (BMFSFJ, 2005b).

Last but not least, Blinkert and Klie (2000) found evidence for the fact that social class is also an important indicator for the willingness to provide family care. For Germany – and this might be also true for Britain – they came to the conclusion that this willingness is highest among persons with a traditional working-class background. As an explanation for this result, Blinkert and Klie refer to the opportunity costs which result from the care of a family member: these costs vary as between different social classes, being relatively low in the working class.

Non-kin relationships

One of the major changes in the study of intergenerational relations over the last twenty years has been the acknowledgement of the role of non-kin. For those living alone without family, non-kin in the form of friends and neighbours play a crucial role in both countries (Phillipson *et al.*, 2000; Reichert *et al.*, 2003). However, the support from friends and neighbours is often of a different kind to that from families: whereas friends mainly provide emotional support, neighbours help with instrumental tasks of everyday life such as dealing with mail and packages, picking up groceries or 'checking' on the nearby elderly resident.

In a study of three areas of England, friends were crucial in the support networks of middle-class older people, particularly in relation to confiding, and especially if they were childless and lacked intergenerational potential. More than half of the older people in the English Longitudinal Study of Ageing (ELSA) (Institute for Fiscal Studies, 2002) said they met up with friends at least once a week. Equivalent results can be found for Germany (Minnemann and Lehr, 1994; Wagner *et al.*, 1996). Friends, however, tend to be those within a peer group and of roughly the same age. The emergence of 'fictive kin' of different generations is a contemporary feature: this is a term used to refer to individuals, unrelated by either birth or marriage, who have an emotionally significant relationship with another individual which would take on the characteristics of a family relationship. Such examples would be godparents or old family friends whom children call 'aunt' and 'uncle'. We know very little about their contribution to and from older people.

In all, friendships are important elements in quality of life (Godfrey *et al.*, 2004; Reichert and Weidekamp-Maicher, 2005). The absence of social relationships, or a low level of engagement, are likely to have a detrimental influence on quality of life, health status and well-being (Lehr and Thomae, 1987). The study by Godfrey *et al.* (2004) also showed that central to quality of life was the value attached to interdependence and being part of a community where reciprocal help and support were available. From Germany we get equivalent results: older people here not only place great emphasis on emotional assistance and meaningful connections (Ferring and Filipp, 1999) but also on the maintenance of 'threatened' values such as reciprocity, autonomy and self-sufficiency (Minnemann, 1994).

Despite different household configurations, neighbours also provide valuable support and play key roles in Britain and Germany in terms of providing intensive help as well as coordination of services. In some cases neighbours even become care givers who provide personal, even intimate care, although – as a comparative study between England and Germany shows – many neighbours and friends see limits when it comes to non-kin care giving. For example, more German than English non-kin care givers are unwilling to provide personal care or care for an older person who suffers from dementia (Nocon and Pearson, 2000; Reichert *et al.*, 2003).

The importance of friends and neighbours in the contemporary life of older people has led some commentators (Phillipson *et al.*, 2000; Pahl, 2002) to comment that relationships are much more 'voluntaristic', giving people choice in their relationships. The implications for social policy in fostering such roles are critical if people are to have a choice of supports in later life. The challenge for the future therefore is to support intragenerational as well as intergenerational relationships (see below).

The macro level

Welfare regime in Britain and Germany

In order to examine the intergenerational relationship on a macro level in Britain and in Germany it is necessary to have a look at the welfare regimes in both countries. In distinguishing welfare states one of the most used typologies is that of Esping-Andersen (1999), with Britain being 'market liberal' and Germany 'conservative-corporate'. His description provides a useful basis for comparison between social security systems, i.e. Beveridge and Bismarckian welfare regimes (see this volume, Chapter 2).

In the last ten years both Germany and Britain have undergone welfare state reform. Since 1997 Britain has developed a New Labour agenda of modernising welfare, with changes in the pension system, health and social care arrangements and funding of welfare. This has taken place alongside devolution of government and responsibility for health and social care to Wales and Scotland. In Germany – mainly as a reaction to the demographic development – there have also been a lot of changes in the social security system and a strong cut in the social welfare budget.

Against this background there has been a fundamental public discussion about the consequences of these changes and the shortage of resources – e.g., in the health care sector – for the relationship between generations (Schmähl, 2002). At one extreme the opinion is held that, in the long term, the younger generation will have no opportunities to shape things themselves, either in political or economic terms, since elderly people will be living at the expense of the young and will deprive them of their prospects (Gronemeyer, 1997). According to this view, the resulting conflict between old and young is determined by the fact that, in the experience of the young, older people, as the ones now ruling, have used up everything that should have served as the basis of life in the future for the young. As a consequence a 'war of generations' or the dissolution of the 'intergenerational contract' seems possible. Attias-Donfut and Arber (2000), who have – as mentioned already – a more optimistic view, remind us that such a contract may be threatened in the future by the 'rolling back of the welfare state', particularly where reform leads to changes in gender relations.

Definition of 'intergenerational contract'

The term 'intergenerational contract' has been adopted to indicate the mutual, and in particular, material dependency and obligations of the different generations. It describes the agreement, either based on social values and norms or determined by law, whereby the middle generation provides financially both for children who are not yet in work and for elderly people who are no longer in work, in return for which they may hope for a similar provision to be made for them in old age. Although the intergenerational contract has traditionally been developed on an individual model it is crucial that this links 'generations'. Therefore, the modern form of intergenerational contract is a redistribution between social generations in the life cycle, a good example of which is the financing of pension schemes. The relationship between age groups in society depends on the extent to which intergenerational solidarity is actually practised and recognised as a social norm. However, as mentioned already, the lack of resources within welfare states and the challenges for states and families to balance care between them brings the issue of the intergenerational contract to the fore. Therefore in Britain and in Germany there has been a discussion in recent years on how the solidarity of the generations can be safeguarded in future at times of demographic change and changing family structures. Or in other words: whether or not the so-called intergenerational contract will work in the future.

Intergenerational relationships and social security

In the light of a rising proportion of older people in the population as a whole, there is increasing talk of older people being an economic burden on younger people, which is reflected in terms like 'the burden of old age'. Pensions and saving for future generations of older people is a current topic of debate, one which can be seen to lead to potential conflict between generations. The same is true for the discussion on the costs of health and social care: the growing number of older persons is held responsible for what is called the 'costs explosion' in this sector. In Germany this has led not only to a rise in health insurance contributions but also to significant cuts in benefits. In Britain older people are accused of 'bed blocking' in hospitals.

In this connection, the dependency ratio of people older than 65 years to those aged 20-64 years old is mentioned. It will increase from 27 per cent in 2004 to 48 per cent in Britain and to 55 per cent in Germany by 2050 (Pensions Commission website for the UK; DeStatis website for Germany). Therefore, the ambivalence discussed earlier on the micro level of intergenerational relationships continues on the macro level. Here too conflicts can arise in which the legitimacy of established distribution rations between generations is called into question (Motel-Klingebiel and Tesch-Römer, 2004).

However, quantitative shifts in the relationship between age groups do not automatically mean that the intergenerational contract is in danger. Firstly, only if the generation capable of gainful employment were to assume that the younger generation was going to denounce the intergenerational contract would this lead to a major conflict between generations (Naegele and Schmidt, 1998). Secondly, it should be kept in mind that the problem of distribution in the welfare state cannot be attributed to demographic change alone (Naegele and Schmidt, 1998; Schmähl, 2002). Thirdly, very often the political debate on economic activity is limited to employment. Elderly people are also economically active in many different ways:

- They add value by making financial resources available which they have accumulated in the course of their lives, and which can be used to finance investments.
- They carry out important work for their family members (see above).
- They are consumers.
- They participate in the financing of public services, e.g., of schools, higher education institutions, etc. through their tax payments.

In addition, sometimes there is a failure to recognise that as they pass through the different phases of life, people are at times net contributors and at times net beneficiaries (Schmähl, 2002). Throughout, it is important to consider a life course perspective, which links all the phases of life. As already mentioned, although this has traditionally been developed on an individual model it is crucial that it links generations. If the ambivalence is not to dissolve into a breakdown of solidarity between generations it is the particular task of politics to prevent insecurity on the part of the younger age groups, for example with regard to old-age pensions, and to strengthen the confidence of the 20- to 50-year-olds in the lasting validity of the solidarity pact (Rürup, 1999).

Social policy debate

In this sub-section we turn our attention to the policy debates in both countries. Comparisons are drawn between the two in relation to policy surrounding the family versus state responsibility for welfare of older people and policies to strengthen solidarity. The debates are played out by way of the examples of the concept of independence in community care policy for older people, the work–life balance agenda, and pensions. These have been chosen because of the similar level of debate in the two countries. Age discrimination and social inclusion are debates in Britain which are yet to gain a profile in Germany.

Care and independence

'Independence' has been the mantra that has been followed through a combination of 'community care' policies in both countries. This has been supported through recent policies such as pension credits, direct payments, legislation to assist carers, and long-term care insurance, alongside privatisation and a modernisation agenda in Britain and in Germany over the last ten years. The concept of independence, however, has consequences for both the family and intergenerational issues of solidarity. As we mentioned at the beginning of the chapter, gender has a significant role to play in maintaining generational and family solidarity, which can be threatened by changes at state level to benefits and long-term care designed to foster the concept of independence. The current debate on the funding for long-term care, (detailed in this volume, Chapter 9) has consequences for the family. In England, means-tested social care, targeting and rationing of resources have left many older people totally reliant on family support for low-level preventative services (Clark *et al.*, 1998) as well as showing high levels of dependency. In Germany, the introduction of long-term-care insurance indeed gives more independence to the frail older person to choose between institutional care, formal community-based services, a cash payment, or a combination of the latter two. Still, with its aim of promoting (informal) community care instead of institutional care and by paying a cash allowance that the frail older person can use as he or she wishes, long-term-care insurance also puts pressure on families to care, especially on women. Thus, one might argue, the freedom of choice and independence of the latter is restricted, or in other words, they feel a stronger commitment to care for family members from which it is hard to escape. This opinion is favoured by feminists who argue that the core problem of long-term care in the community is the exploitation of women. In essence this has meant dependence on women in the family, making a mockery of the notion of independence for older people and the relatives who support them. The concept of 'interdependence' is missing from policy debates. However, this has not led to a breakdown in solidarity and the government, conscious that this needs to be strengthened, has introduced a number of initiatives to support the growing number of women who combine work with care of older people.

Work–life balance

The 'Work–Life Balance' initiative was introduced by the British Government in March 2000 to help employed carers, it included a new £1.5m Challenge Fund to help employers explore the work–life balance. This has been a crucial step in assisting carers to juggle work and family life, and builds on a Carers Strategy and a raft of legislation (for example, the Carers and Disabled Persons Act 2000). The debate has centred around the case for supporting carers looking after either older people or children. Despite the 'sandwich generation' (i.e., those 'caught in the middle' between parent care and child care) being an atypical experience at present (Evandrou and Glaser, 2004), for future cohorts this could be a realistic prospect. The implications for intergenerational solidarity and conflict may yet emerge unless policy takes a life course approach and supports those women who have multiple breaks in their working lives to care for relatives. This counts especially for Germany: here the socio-political discussion on 'work–life balance' still revolves round combining work and child care. Although specific research on work and elder care was conducted as early as 1995, hardly any effort is being made by politicians, companies or trade unions to introduce measures which support employment and care for *older* people (Reichert and Naegele, 1999). The only outcome from the research project – which was funded by the then Federal Ministry for Family and Senior Affairs – was a brochure which had the aim of informing important actors on the subject.

Pensions

In Britain there is a need to move beyond the voluntaristic approach of today's pension system. This is heightened by the fact that intergenerational transfers of wealth are less likely to occur as current older consumers prefer to spend their money (see this volume, Chaper 4). Initiatives such as pension credits and winter fuel payments have been introduced. However, unless people work longer than they are doing currently, significant numbers of today's workers will not build up the level of retirement income they might expect. There is also a particular issue for women. The government believes that people should take more individual responsibility for their own security in later life and wishes for a consensus behind this.

This is also true for the German government. To face future challenges Germany's pension policy has undergone and is undergoing different reforms (Bäcker *et al.*, 2000). All these reforms have to be seen against the background of the main socio-political goals, of which the most important is the adjustment of the social security systems which are based on the social insurance principle and financed by the pay-as-you-go system in response to demographic changes. Social policy wants to ensure the continuance of the intergenerational contract and justice between generations by measures like modifying the formula for calculating pensions, by introducing a factor which aims at balancing the relationship between the number of contributors on the one hand and the number of pension recipients on the other hand. As one result, German workers need to prolong the period of paying contributions in order to receive a sufficient pension level. Additionally, taxation of high pensions and other forms of old-age income sources such as interest or rent, and the raising of the statutory retirement ages, have been or will be introduced (see this volume, Chapter 2).

Age discrimination

This is a further policy dimension that affects generational solidarity. There are a number of policy debates, which are different in Germany and Britain. In Britain age discrimination, age equality and social inclusion are key political issues where the profile of older people has been raised. The role of older people as citizens and consumers is highlighted in recent legislative moves to outlaw discrimination in the workplace. Additionally an Age Positive campaign through the Department for Work and Pensions promotes the benefits of employing a mixed-age workforce that includes older and younger people. There is a necessity, however, to go beyond just workplace-based discrimination. For example, conflicts in terms of health rationing based on age need to be considered. This has taken the form of debates in the health and social care service about 'bed blocking', where older people themselves, rather than the system (of funding long-term care and of arrangements for domiciliary care), are perceived as the problem at the interface of heath and social care. In Wales the establishment in 2006 of a Commissioner for Older People aims to tackle such issues, yet a major debate around the image of older people and the discrimination they face in health and social care has to be of wider concern.

Compared to Britain the discussion on age discrimination in Germany has lagged behind. Although age discrimination takes places in everyday life (older people not only have problems at the workplace but, for example, also have difficulties in getting a loan from a bank or joining a private health insurance scheme), an anti-discrimination law, as required by the European Union, had not been passed by the German parliament. The proposal stalled in a debate over implementation details. Moreover, the parties involved could not agree how far beyond basic EU requirements the law should go. However, the law, called the *Gleichbehandlungsgesetz* (equal opportunity law), was passed and came into force in 2006. This measure outlaws discrimination not only on the basis of race, religion, disability and gender but also on the basis of age.

Future challenges

To make sure that there will not be 'a war of generations' – either on the micro or on the macro level – the British and German societies are confronted with different challenges. Some of the most important ones are outlined below.

New definition of 'family'

There is no doubt that the definitions of 'family' and 'generation' are changing. In Britain and in Germany we not only have the nuclear family. Instead, we observe that families are becoming more complex and can take on many different forms (e.g., single mothers and fathers, patchwork-families, family members 'living apart together', four- and five-generation families, divorced spouses who are still committed to each other). Social policy in both countries has to react to this development because such complexities have challenged the notion of a family contract. However, it has to be noted that the effects of policy in bringing order out of complexity can help or hinder families. 'Diverse' does not mean dysfunctional. Changing family structure does not determine family function. Although evidence suggests that relationships within families are (still) strong, the notion that older generations will certainly be cared for by younger members of their families is no longer guaranteed. In British social policy, traditional responsibilities of different generations within the family in relation to care have shifted with the introduction of direct payments, giving those requiring care the right to choose what kind of care they prefer and who should provide it. German long-term-care insurance also took into consideration that it might not be family members who want or who are able to provide care for an older person. As just mentioned, the care recipient also has the freedom of choice whether he or she wants to be cared for by family members, by friends and neighbours and/or by community services. However, social policy has to find more innovative ways to strengthen inter- and intra-generational solidarity because 'care for older people' is just one of many fields of generational solidarity. For example, older people support the younger generation and also their own generation within and outside the family in many different ways (see this volume, Chapter 5). This fact is not considered sufficiently by social policy and by young people. Also, the increasing use of assistive technology in reproduction also poses particular issues for defining the family and for assessing the potential of generational transfers.

Role of social services and the state

From the OASIS study (Lowenstein and Ogg, 2003) mentioned earlier, with its comparison between Britain and Germany, it also appears that policy should support a redefinition of the role of the family in care provision, allowing a mix of informal family care and formal service provision, but with the state in a more central role than at present.

The study found that access to services increased their use and was welcomed by all generations, with elderly people rather more reluctant to receive family help than adult children were to provide it. Receiving help from the formal sector enabled elderly people to maintain their independence and autonomy. Services should, therefore, be more accessible to older people. Because of the interconnections between the formal and informal sectors, the ambivalence which – as already mentioned – also characterises family relationships has to be addressed by social services. It will therefore be especially important in the future to organise formal social services and their financing in such a way that they form a framework conducive to stabilising the relations between the generations based on solidarity. It also has to be kept in mind that social services are used by different generations, i.e., also by the very young generation of children and adolescents. Although we can expect – due to the low birth rate – a decreasing need for services for children and adolescents, this development has to be accompanied by a qualitative increase and greater differentiation of requirements in particular.

Minority ethnic families

The increasing numbers of minority ethnic older people also challenge the concept of the intergenerational contract. In areas of Britain there is still evidence of minority groups, particularly Bangladeshis, living in multigenerational households with different generations supporting each other (Phillipson *et al.*, 2000a). For other ethnic groups, e.g., Chinese communities, different patterns of intergenerational relations can be found, contrary to the common view of a lesser capacity in Chinese families to meet the needs of older members (Chau and Yu, 2000). In Germany many migrant families are also afflicted – although slowly – by changes in family relationships and family-oriented values and norms which endanger the traditionally strong intergenerational solidarity (BMFSFJ, 2001). In consequence, social policy in Britain and in Germany has to focus on migrant families, too. For example, social services will have to face the challenge of providing differentiated assistance in the integration of migrants.

New definition of the (inter)generational contract

An important challenge in the future is also defining what sort of (inter)generational contract we need and what sort of social policy we need to support it. The younger generation has many concerns (e.g., the financial burden of old-age pension systems, damaged environmental resources, not enough opportunities for political expression) which have to be taken seriously by policy. We need a policy which does not pass on the pressing problems of today to subsequent generations. Rather, policy has to be guided by the principle of intergenerational equity, i.e., no generation should be deliberately disadvantaged by another.

Solidarity with the younger generation and one's own generation outside the family

The change in intergenerational relationships within families, and the move away from traditional norms, means that in the future relationships outside families – especially in old age – will become increasingly important. Such relations can be fostered partly trough voluntary work. Voluntary work of older people for younger people and vice versa, and for their own generation, can be regarded as a supplementary resource, for example, for social services, and this also applies to the exchange between generations.

The diversity of older people

Social policy has increasingly to recognise and react to the many different groups of older people (e.g., single women without children living together, gay and lesbian relationships, older people with learning disabilities, older people with no or only fragile social networks), and to their different life situations, needs and wishes. Whether or not all these groups of older people receive support and care will depend on the future definition of 'family', because this in turn will influence the political actions and measures available to them. Further, the facilitation of inter- and intragenerational support outside the family has to become an important socio-political goal (see above). However, the engagement of every individual in doing something for someone else is asked for in this context, too.

The potential of older people

It should be mentioned again that it will become necessary to use the potential and competencies of older people, not only for voluntary work but also in general. Never before have older people been on average in such good physical, psychological and intellectual condition, as materially secure and as active as they are today (Naegele and Rohleder, 2001). These increased resources and skills of older people, and their considerable amount of spare time, have led to a debate as to how they should be put to use in society. However, older people today already make a substantial contribution towards the cohesion of society. Additionally, considerations of productivity in old age must be matched to the many different life situations in old age. For example, many older people cannot be, or only to a limited extent, socially productive for health reasons. Therefore, the socio-political framework conditions have to satisfy the needs of both the individual and the society in equal measures.

In addition to these challenges we would like to mention the following ones:

- Supporting women to juggle work and family life and to maintain a role as kin keepers without overload and defining the role of men in this context; setting up measures to make it more possible to combine work and family life presents a special challenge for politicians, but also for industry.
- It will be important to avoid generational war over issues such as financial support in retirement, health rationing, etc.
- It will be important to maintain the productivity of younger older people, particularly in Germany where an economic downturn can leave older people out of the labour market and can lead to resentment.
- It will be important to take a holistic approach to policy; in housing, for example, ensuring balanced communities rather than age-segregated housing developments.

Lessons for each country

In Britain as in Germany, over the last ten years intergenerational practice has emerged as a research area. However, the synthesis between the two research communities of intergenerational relationship research and intergenerational practice research is missing. One of the challenges is to ensure an intersection between the two arenas, particularly if best practice is to be highlighted. In this section we give one example of 'best practice' from each country as regards how inter- and also intra-generational relationships can be improved in an attempt to achieve this process of synergy. We can learn lessons from these practice examples in order to strengthen generational ties in both countries.

Our first example refers to the micro level of intergenerational relationship and comes from Britain. 'Mentor: Adfam and Grandparents Plus Project' is a project which aims at helping grandparents who have adult children with serious drug problems. Due to their drug problems these adult children are very often unable to bring up their own children – a task which in many cases has to be fulfilled by the grandparents. However, grandparents often lack knowledge about their rights and responsibilities, have access to very little information, are unaware of sources of help or support, and often want guidance about the day-to-day practicalities of living with children / young people (Richards, 2001). On the other hand, research points to the benefits for young children who are raised by grandparents, as opposed to being in foster care or children's homes (Broad, 2000).

The Mentor Project works with 12 grandparents who are raising their grandchildren and who are also actively involved with community organisations which support grandparents. In a first step these grandparents help to identify what information, support and advice grandparents raising their grandchildren need. In a second step the project aims at providing these grandparents with training and support to develop their skills and knowledge about drug and alcohol use and how to prevent drug-related harm to young children. The idea behind this procedure is that these 'educated' grandparents are to act as a resource for other grandparents raising or caring for grandchildren in relation to substance misuse issues. They will in turn disseminate what they have learned through their formal and informal networks. In other words, the Mentor Project not only stabilises intergenerational family relationships but also has a focus on intragenerational self-help and support – good reasons for us to choose it as a project which should be transferred to other countries such as Germany.

In Germany the policy debate focuses on intergenerational practice, and projects to find good models outside the family. This is also true of the project called *'Dialog der Generationen als interkulturelles Projekt'* (Dialogue of Generations as an Intercultural Project). The main aim is to improve understanding between young migrants and old Germans outside the family context and to minimise the prejudice each group has toward the other. Young migrants visit older people who live in the same district as themselves and help them with everyday tasks like shopping or household chores. However, these kinds of support are only a peg to bring the two groups into contact, and make them talk to each other with the final aim that they become friends. In preparation for their 'job', the young migrants get basic information on how to help older people with special needs (e.g., how to push a wheelchair) and how to behave in conflict situations. The young migrants get a small expense allowance for their help and they can visit the older person they have 'adopted' as often as both 'partners' like. What makes this project so special is that it not only fosters the dialogue of generations but also brings together people from different ethic backgrounds. In this connection it is important to know that the main reason for initiating this project was a politically motivated right-wing extremist fire attack against a home for migrants.

Globalisation makes it even more important that Britain and Germany share their experiences, knowledge and policy. Similar problems appear in both countries and both countries have to answer the challenge. But Britain and Germany need more: they need a policy which is sensitive to generational issues in general.

8
Health Status and Health Policy

Anita Pfaff

Introduction

With increasing age, health and the health-related quality of life tend to decline. This process is especially pronounced beyond an age of around 60 years. It manifests itself in the form of a higher incidence and prevalence of certain diseases and also through more severe symptoms of some of the diseases which occur in other age groups as well:

- Certain physiological changes occur predominantly in advanced age (e.g., arteriosclerosis, osteoporosis).
- Diseases with long periods of pre-clinical latency are more prevalent in older age (e.g., different types of cancer).
- Some diseases show different manifestations in advanced age (e.g., infections).
- The incidence and the prevalence of chronic diseases of long duration or repeated exposition (e.g., of the respiratory system, diabetes type II, high blood pressure) increase with age (SVRKAiG, 1996; Kruse *et al.*, 2002, p. 10).

The ageing process is sometimes accompanied by a gradual or at times more accelerated deterioration of health. Even though a marked decline in health does not occur at the same chronological age for everybody, for most adults their health care needs start to increase as their health-related human capital 'depreciates'. (Another age group with higher health care needs, however, are infants. We therefore speak of U-shaped health care needs and associated health care expenditure with regard to age differences.) Even though preventive efforts on the part of individuals and on the part of society will slow down this process,

165

adequate health care becomes more important with increasing age for the majority of people. In order to enable them to maintain an adequate health-related quality of life, on average more economic resources have to be spent for the health care of an older individual than for a younger one. In all industrialised countries this involves major public spending as well, even though the public–private mix may vary considerably between countries (OECD, 2006).

In all countries, especially the industrialised ones, these age-related health differences have been accompanied by a long-term trend of improvements in health which have occurred in all age groups over the last century. These changes have not only led to a decline in mortality – most markedly, though not exclusively, a decline in infant mortality – but also to a drastically reduced morbidity for most groups. Improvements in social and economic conditions were major determinants of this development. But similarly advances in medical technology and knowledge as well as the development of public health care regimes, which made health care accessible for the population at large, have been responsible for these improvements. During the last few decades, moreover, some improvements in health and health-determined quality of life can be attributed to the application of new methods of diagnosis and treatment of diseases (e.g., by-pass operations, medication, knee- or hip-replacements).

Two conflicting theories have been advanced to explain differing observed effects of the increased life expectancy and the resulting health care needs and health care expenditure: the compression of morbidity (Fries, 1980; Fries *et al.*, 1989; Fries, 2003) and the expansion of morbidity (Olshansky *et al.*, 1991) theories. While in principle medical knowledge – at least at a price – is available to all, for all practical purposes social, political and economic circumstances determine access to health care. These conditions differ between countries. Even in rich countries health care is not equally accessible and available to all groups. Social justice and equity have not been fully achieved in the UK and in Germany, notwithstanding their comprehensive public health care systems. For example, *The Black Report*, commissioned by the Labour government in the late 1970s to investigate possible inequities of the health care system, produced rather shocking results, which were, however, played down by the then new Conservative government to which it was presented (Department of Health and Social Services, 1980; Townsend *et al.*, 1992). Germany also has great differences in the health status of different social groups (Mielck, 2000). Different kinds of social inequities, however, persist in all wealthy, industrialised countries (van Doorslaer *et al.*, 2004, pp. 8–9). Most OECD countries claim equity objectives for their health care systems. However, there are some discussions as to whether specific equity objectives should entail the effort to achieve an equality of outcome, or the less ambitious objective of equal accessibility (Hurst and Jee-Hughes, 2001).

Even though the stated general policies of different countries may sound very similar, the material outcomes and even the concrete form of access to goods and services may differ. Marked differences in historical development, the philosophy, objectives, organisation and the general level of generosity of public health care can be observed across countries. A general consensus exists that the equity objective implies horizontal equity and equal access for everyone. Nonetheless, this allows for a wide variety of interpretations of the term 'accessibility' (Goddard and Smith, 2001).

Health care today is of special importance to older people, since they are – statistically speaking – the major users of health care services. Their declining health requires an increasing amount of goods and services, in order to maintain a reasonable quality of life up to an advanced age. The increasing technological know-how which has contributed to an increase in life expectancy and reduced morbidity has resulted in resource use and expenditure patterns which make persons past retirement age, especially the very old, the main users and beneficiaries of health care goods and services. This was not always the case.

Previously, the major premature loss of life was caused by infectious diseases. And a very high mortality rate occurred during birth and the first year of life. (Old) public health, better hygiene and understanding the causes of infections were the main reasons for improvements in population health and the increase in life expectancy. Today many of these diseases play a minor role, since they can be contained through prevention (inoculations, vaccinations, improved nutrition, living and working conditions) and treatment (e.g., with antibiotics). The most prevalent causes of death in high-income countries today are ischaemic heart disease, cerebrovascular diseases, diseases of the trachea, bronchus and lung cancer, diabetes mellitus, chronic obstructive pulmonary disease, lower respiratory infections, Alzheimer's and other dementias, colon, rectum, stomach and prostate cancer (Mather and Loncar, 2006, p. 2023). And the major nutrition-related diseases today in industrialised countries – especially among the low-income groups – are overweight and obesity and further diseases resulting from this disorder, rather than from malnutrition and hunger. In this respect Germany has a poorer record than Britain.

The UK and Germany have instituted and developed very different social and health care regimes for their citizens (Esping-Andersen, 1990; Döring *et al.*, 2005). It is therefore of interest to compare the performance of the two systems with regard to health care for older people, even though the differences in health care regimes as such cannot be held solely, or even primarily, responsible for differences in health outcomes. This is especially true for the old, since their health status is the result of a great number of different influences over their entire life course.

The health care regimes, of course, are not confined to dealing with older people. However, their overall characteristics have to be considered, in order to identify their specific impact on this group. Even though the UK and Germany have very different health care systems, the two countries have certain problems in common and certain reform trends follow a similar course. At the same time the two countries differ in some aspects with regard to specific goals, policies and outcomes.

The UK and Germany – Beveridge vs Bismarck

The UK and Germany as members of the EU and OECD are both highly industrialised wealthy countries. Their general health care objectives do not contradict each other: today they share common norms (and also with many other industrialised nations), such as seeking a more or less universal coverage of their populations/residents against needs arising from social risks, specifically from health hazards and their consequences. This implies that the countries have to ensure that the necessary and required health care services and goods of high quality are made available efficiently to 'everyone' – independent of an individual's ability to pay. In the pursuit of this general objective, however, the prevailing constraints have to be taken into account. For historic and political reasons the two countries have taken different paths in developing collective health care systems. The two countries have also chosen somewhat different concrete specifications when it came to qualifying and quantifying general terms such as 'everybody', 'necessary', 'needed', or 'high quality', or when and how they define the services to be provided. Furthermore, the equity concept which is implicit in the health care objectives can be defined in more than one way. In the context of this chapter we are, of course, primarily interested in how the general principles affect older persons, or whether older persons are subject to positive or negative discrimination.

The UK's Beveridge-type system

The UK's Beveridge system is based on the National Health Service (NHS) of England, Wales, Scotland and Northern Ireland. In fact the UK does not have *one* uniform health care system, since, especially in the course of devolution, the health care systems of England, Wales, Scotland and Northern Ireland have taken slightly different courses. Our main emphasis will be placed on the English health care system. We shall often speak of the UK, though, because despite the differences, for example in their implementation of reforms, the four systems have many basic features in common. The English NHS states as its ambitious founding principles to provide 'quality care that:

- meets the needs of everyone;
- is free at the point of need; and
- is based on a patient's clinical need not their ability to pay (NHS England, 2000).

These principles ensure a health care system characterised by coverage of 'everyone's' health care needs. This implies a statutory universal coverage – as opposed to coverage of only part of the population or a differing coverage for different population groups. Compared to the inadequate situation prior to the founding of the NHS in 1948, this was a great leap forward and not surprisingly, it took some years to implement.

The second principle implies two elements. Firstly, it follows that providers are not directly reimbursed for their services by patients at the point of delivery and, secondly, that in principle no co-payments are to be charged. This important characteristic frees patients of the obligation to pay the necessary price at the point of delivery, when they are faced by the (possibly acute) need for health care goods or services, but do not have the money to pay for them. It is thus a prerequisite for the third principle, which states that the provision should depend only on individual need, not an individual's ability to pay.

At first sight these principles suggest a very generous system which does not face scarcity of resources. However, this system – like every other one – has to operate within the economic constraints of limited budgets. While a *pure* market would allow demand (backed by the ability to pay) and supply (governed by cost constraints) to regulate an excess demand over supply through adequate pricing, it would thereby violate the equity principle of social justice (allocation according to individual need), replacing it by allocation according to the individual's ability and willingness to pay. In principle, allocation according to need alone could imply infinite demand. The NHS needed to constrain an excess provision of goods by some other means. These took account of the existing need. But, in the face of rapidly expanding medical know-how, they constrained the benefits to a level and capacity which did not make every possible diagnostic and therapeutic measure immediately available to everyone. This resulted, for example, in long waiting-lists or in *de facto* rationing.

The services of the NHS, which did not involve payments at the point of delivery, required financing. Since necessary health care is viewed as a public good, this logically implied financing out of tax revenues. (The wide range of available goods and services which partially satisfy health care needs, but partially can be considered to be lifestyle products, has to lead us to question the general public-good character of specific goods and services. Some have very strong private-goods characteristics which do not warrant free availability. In practice, it may be difficult to determine whether a specific good or service is a necessary health care good or a general consumer good, e.g., health food, anti-ageing or 'wellness' products.) As is well known, decisions on the allocation of tax revenues are governed to a large extent by (often short-term) political considerations – and not necessarily by the individual needs of the population in one area. Over the years the NHS, like all other national health care systems, had to undergo many reforms and adjustments, necessitated by basically similar problems: aspects of equity, quality of care, efficiency of delivery, technological advance and patients' rights have needed adjustments over the 60 years of the NHS's existence. Major reforms have occurred since the early 1990s, first with the introduction of quasi-market elements and partially with additional private funding. This development continued under the Labour government from 1997, leading to the radical reform laid down in the formulation of *The NHS Plan* of 2000 (Department of Health, 2000), which constitutes the biggest change to health care in England since 1948. It sets out a ten-year plan to modernise the NHS through organisational reforms and markedly increased funding. Within five years the funding was to be increased in real terms by one-third. In fact the NHS budget almost trebled from 1997 to 2008. The main objectives for the ten years to follow included 7,000 additional beds in hospitals and intermediate care, the establishment of 100 new hospitals and 500 new one-stop primary care centres, and a massive expansion of medical personnel. In addition, the philosophy and organisation of the NHS was to be changed – focusing on patients' needs and preferences.

A large number of steps have been taken to realise these ambitious plans. These are of special relevance to the older people who are the main users of the NHS. For them the National Service Framework for Older People set up ten-year targets in relation to the implementation of *The NHS Plan* (Department of Health, 2001b).

The three general principles outlined above continue to govern the NHS. However, they have been differentiated and modernised in the context of the health reforms in line with the reform objectives in the wake of *The NHS Plan*, emphasising clinical excellence and patients' autonomy and efficiency rather than the initially more paternalistic philosophy of the NHS.

The NHS continues to distinguish between primary and secondary care (Figure 8.1). Patients have to register with a general practitioner (GP) whose services they can then use without direct payments. The GPs or 'family doctors' continue to exercise the role of gate keepers and stewards since they normally refer their patients to secondary care, such as hospitals and specialists, and they prescribe medication. Dental services have been privatised more and more during the last few years.

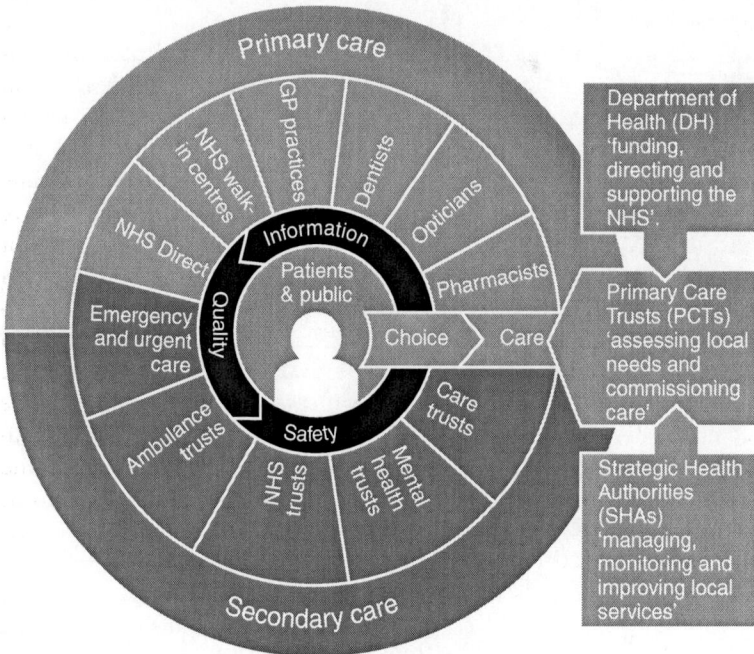

Source: www.nhs.uk/england/AboutTheNhs/Default.cmsx.

Figure 8.1   The organisation of the National Health Service in England

The recent reforms have also emphasised a shift towards 'modernity' by offering additional services like NHS Direct, a telephone consultancy, and walk-in centres. Moreover, the serious shortages and shortcomings, which of course had also contributed towards making the UK's health care system a very thrifty one, have been tackled through such modern instruments as more patient orientation and transparency (e.g., advice and information about waiting-lists) and stricter quality control (e.g., via the establishment of the National Institute for Clinical Excellence (NICE) in 1999). Patients' choice, which is always somewhat limited in a primary care system, was extended, especially once patients had had to go on waiting-lists of six months for an elective surgical procedure.

In order to gain more transparency with regard to purchasing and providing, (local) Primary Care Trusts (PCTs) have been established starting in 2002. They control the major part of NHS funds, assess health needs and commission care. The allocation of funds to PCTs is based on weighted capitation targets. In order to overcome the resource shortages of the past, funding allocations have been increased considerably over the last few years (Department of Health, 2005b).

The concern with past inefficiencies of the system, partly caused by the combination of financing with the provision of services, led to the establishment of NHS Trusts for hospitals in the early 1990s which made it possible to separate the roles of purchaser and provider. While these trusts were in principle to be given managerial competencies, they remained accountable to the Secretary of State for Health.

Since 2002 the efficient and successful trusts (hospitals) have been allowed an 'earned autonomy' by being transformed into NHS Foundation Trusts (also called NHS Foundation Hospitals). These NHS Foundation Trusts are independent public-interest companies intended to exercise managerial freedom (Robinson, 2002, pp. 506–7; Klein, 2004, p. 1332). They may, for example, determine their own investment plans or retain revenues from land sales. By transferring accountability to local organisations, it is hoped that efficiency will be enhanced and the integration of health and social services will be improved. This transformation is still in progress (Department of Health, 2004; Busse *et al.*, 2005 pp. 59–60).

Germany's Bismarck-type-system

Germany's public health care system was developed much earlier than the NHS. Actually, in the early and mid 19th century some forms of public or mutual health care efforts developed in some principalities that became part of the re-established German Empire in 1871. However, usually the founding of German Public Health Insurance in 1883, providing coverage against the health hazards of the industrial labourers for the entire re-established German Empire, is considered to be the beginning of a public health care system in Germany (Lampert and Althammer, 2005; Bundesministerium für Arbeit und Soziale Sicherung, 2006, pp. 126–32). Together with Industrial Accident Insurance it was one of the first social insurance schemes of the 'Bismarck system'. Unlike the UK's universal system, it originally provided coverage for a relatively small group: industrial blue-collar workers. It was typical of the early social insurance system to be based on separate solidarity arrangements of different social groups. The industrial workers were *one* group that seemed in particular need of health care. One very important feature of this system was the compulsory nature of the insurance for industrial workers, especially since the employers had to share in its financing.

The major benefit provided at the initial stages was wage replacement income – sick pay – for sick workers. Beyond that, doctors' services and medication as well as maternity benefits were provided. The financing of services initiated a shift from wage replacement towards health care goods and services. Individual health funds could also include benefits for dependent family members.

Coverage was expanded to further social groups gradually over the decades. The most important reforms included further groups of employed workers in 1914 and the recipients of Public Old Age Pensions in 1941. For older people this constituted a most important step forward, slowly changing the system in the direction of universal coverage. This process took longer in the Public Health Insurance than in other social insurance systems. To this day two separate health insurance schemes, governed by very different organisational principles, coexist – Public Health Insurance and private health insurance. Most employed workers as well as recipients of of Public Old Age Insurance pensions are required to be insured in one of the (several hundred) Public Health Insurance funds (*Krankenkassen*); and a free choice of the fund for all members of this group was introduced as late as 1996 in the Health Reform Act of 1992 (GSG, 1992). A small part of the population (especially employed workers whose income exceeds a certain level – €3,975 starting January 2007 – and some of the self-employed) may opt for Public Health Insurance on a voluntary basis – under slightly less favourable conditions than the ones existing for persons with mandatory Public Health Insurance. Alternatively, they may choose a private health insurance scheme. Civil servants – also after retirement – receive a partial reimbursement from their public employer. They are required, however, to buy insurance coverage for the rest of their health care costs. The provisions are such that it generally pays to get private insurance coverage rather than coverage by Public Health Insurance: only if a person is very ill may the risk-determined private insurance contribution be higher than the Public Health Insurance contribution. Private insurance companies may also refuse coverage in such cases. The rest of the population has the option of (voluntary) private insurance. Private health insurance is available at relatively low contribution rates for the young. These rates increase, however, with age – often to a prohibitive level, forcing some persons to cancel their insurance. This has increasingly happened with some (formerly) self-employed persons during recent years.

Only in 2007 did the Health Reform Act Enhancing Competition (GKV-WSG, 2007), a major reform of the Health Care Act (*Fünftes Sozialgesetzbuch* (SGB V)) and other laws governing regulations of the health care system finally succeed in transforming the system into a universal coverage scheme. It introduced the general obligation for everyone to have health insurance by 2009. Some people are unable to afford a normal private insurance scheme, which for older persons or those with high health risks is very expensive. For them a special 'basic' coverage (*Basistarif*) has to be offered by the private insurance companies. This entails an approximately identical coverage to those of the Public Health Insurance funds, at a contribution level which may not exceed the maximum contribution of the Public Health Insurance funds.

These two new measures – the general obligation for everybody to have insurance coverage and the introduction of the 'basic' coverage – are potentially of great relevance to those older people who may not opt into Public Health Insurance. It offers coverage to those who have never had or who had lost their private insurance coverage during a phase of their lives when the need for health care is usually high.

Approximately 85 per cent of the population, i.e., most employees, most retirees and most unemployed persons, have to be (or in some cases, may be) insured in the Public Health Insurance system. It is characterised by more or less standardised, rather generous provision and by contribution rates which are proportionate to wage or wage replacement income, subject to a ceiling level (of €3,562.50 per month starting January 2007). The contribution rates may not be differentiated by current or past health status. The system is financed on a pay-as-you-go basis with risk adjustment between the individual funds. Health care benefits for dependent spouses and children of persons entitled to insurance are covered free of charge. Private health insurance is financed by risk-related contributions. The system is capital-funded, i.e., individual capital accumulations for younger persons are applied to partially offset their higher costs in old age. Despite the capital funding, contribution rates increase with age, since the benefit basket and the prices of health care goods increase considerably over time.

The two parallel systems have different distributional impacts on different age groups. For young singles with higher incomes, who enjoy good health, private insurance offers better options at somewhat lower costs. For most retirees their income is lower than their previous wages. Therefore income-related contributions to Public Health Insurance are much lower than comparable private health insurance contributions. Even though private insurance companies may offer a slightly more generous benefit basket, Public Health Insurance offers better benefits for the aged in relation to the contributions paid. At this stage more than 90 per cent of those of of post-retirement age are insured in the Public Health Insurance system. Therefore older people are affected very deeply by any reform of Public Health Insurance.

Differences and similarities between the health care systems of the UK and Germany

Some differences in the major elements constituting the UK and the German health care systems can be explained by their historical roots. The German system of social insurance was established in the 1880s (Public Health Insurance and Workmen's Insurance in 1883, Invalidity Insurance in 1889). It was largely a consequence of the Industrial Workers' Movement (Arbeiterbewegung), on the one hand, and the Christian Social Ethics Movement (Christliche Soziallehre), on the other.

In the late 19th century the worst consequences of the Industrial Revolution in Germany had become evident. Therefore Chancellor Bismarck attempted to pre-empt the activities of the socialist workers' organizations (Social Democratic Party and trade unions) by introducing social insurance. It was initially confined to providing social benefits to a relatively small part of the labour force, namely to industrial workers. Only a very small economic surplus could be generated beyond the payment of minimal wages and necessary investments, leaving little scope to finance social benefits. Gradually, over the next few decades the scope of coverage – both in terms of persons and in benefits – was extended.

This trend had a double impact on older people. First, retirees were included in health care coverage much later than employed workers and their dependants. Second, including them (as well as their dependent family members) also indicated the increasing importance of financing diagnostic and therapeutic goods and services rather than just wage replacement income for part of the employed population. Today a major part of the expenditure of Public Health Insurance in Germany is directed towards the coverage of diagnosis and treatment of diseases – even for the employed population (Statistisches Bundesamt, 2006a). Furthermore, an increasing share of health care services and expenditure is received by retired persons, i.e., by older persons and especially by persons of advanced age. In a pay-as-you-go system this involves a major redistribution from the working population to those already out of the labour force.

By the time Beveridge (1942) started to design a system of social welfare during the early 1940s, a far more advanced stage of social and economic development had been attained in all the industrialised countries. Furthermore, a larger share of the population lived beyond retirement age (when the Public Old Age Insurance was introduced in Germany in 1889 the retirement age was set at 70 years – an age which very few workers reached at that time). This holds true despite the economic consequences of World War II, which not only decimated the workforce, but drained all economic surplus towards war-related activities. Introducing a new social welfare system following World War II under a socialist government (rather than an anti-socialist one like Bismarck's in the 1880s), it seemed far more logical to try to opt for a universal scheme covering the social risks of *all* citizens – including the old – of the country, especially with regard to the provision of health care services and goods.

Following World War II, Germany had to deal with the reconstruction of its political system, its economy and its social welfare regime. By then it had attained a far more extensive social insurance coverage than during the initial stages – as far as the insured persons were concerned as well as the standard and level of benefits.

The UK, on the other hand, by introducing a radically new social regime, was faced with many transitional problems in setting up a new institution such as the NHS and getting over the World War II-related economic problems. The idea of universal coverage was certainly a visionary approach, but not always easy to implement. However, today, in view of a more extensive harmonisation in the EU context, the universal coverage of the UK system offers a scope which is more future-oriented. This fact has been emphasised by developments in other European countries like Switzerland and Austria which more recently have moved toward universal coverage (Pfaff *et al.*, 2005).

Since around the year 1990 both Germany and the UK have taken decisive steps to reform their health care systems, also integrating some private market principles within the social insurance system and the NHS, respectively, as well as more patient-orientation, more quality assessment, more measures towards increasing their efficiency and more cost-effectiveness.

In primarily market-coordinated economies, market processes are assumed to have priority over public production for the provision and utilisation of goods and services. It can be shown, however, that in the real world market failures and imperfections are likely to occur with regard to the provision of health care (because health and health care cannot be considered to constitute 'normal' private goods) and also with regard to insurance against social risks (because the asymmetry of information leads to negative selection and moral hazard causes insufficient preventive efforts and inefficient decisions) (Breyer *et al.*, 2005). Even market theories therefore call for public intervention in the financing of health care (and more recently long-term care), though not necessarily in public provision and production of (all) services. Public production may, however, be called for in the case of hospital care, where local monopolies may exist.

Germany has certainly had a much larger share of private production and service provision than the UK. The UK adheres more strongly to the understanding that health care is a public good – even in the wake of the more recent reforms outlined above. However, even the NHS purchases medication and drugs from the private sector. Moreover, first hesitant steps have also been taken to offer private franchises on NHS Trusts. In both countries the health care sector is characterised by a high degree of public regulation in those areas where private provision is allowed.

While the NHS's principle of social justice was traditionally realised by the public *provision* of health care according to need, combined with tax financing, the German health care system was for the most part characterised by a separation of purchaser and provider. The German system has been more fragmented. In principle this allowed for elements of competition and (regulated) quasi-market processes more readily, once the strict separation of groups in the social insurance system was overcome. The coexistence of public and private insurance lent itself to limited competition. However, it did not function very efficiently. In the wake of *The NHS Plan*, such elements of private provision and – on account of increased choice – of competition gradually also entered the NHS system. A partial privatisation of social risks took place in both countries – as well as in many others (Busse *et al.*, 2006).

In the face of increasing general wealth and a rising standard of living, of technological advance and of rising life expectancy, the two countries have initially taken different steps in dealing with expanding collectively provided or financed services. The Beveridge-type countries tended to allow for a second tier of voluntary, privately financed coverage (via private health insurance or via direct payments), accompanied by a more strictly budgeted increase in the benefits covered by the NHS. The resulting shortages in capacities have led to dissatisfaction on account of long waiting-lists. Only in recent years has a concerted effort taken place towards expansion of capacities within the NHS. The Bismarck-type regimes, however, have expanded social coverage, by and large, within the scope of Public Health Insurance in a more generous fashion. This has led to a rather rapid, alarming increase in public health care expenditure over the last few decades.

The situation of older people in health care

Health care systems affect all age groups. However, in practice, the national health care systems of the UK and of Germany are organised in such a way as to have different impacts on the older as compared to the younger population. This differential effect is caused, on the one hand, by formally age-specific rules of access (e.g., eligibility for certain diagnostic measures, such as cancer screening, only beyond a certain age) as well as by the *de facto* different treatment of persons of different age (e.g., informal age discrimination).

We shall consider those beyond retirement age, which is normally between 60 and 65 in most countries, as older persons. Among the age group beyond 60 or 65 years two more homogeneous sub-groups are often distinguished. The differences between these two sub-groups are based on their different health status and health care needs: the younger (or active) old are persons up to an age of approximately 75 or 80 years. Beyond that age persons are of advanced old age. The differentiation is more relevant for long-term care than for health care in general. But even in health care a marked deterioration in health can often be observed once persons reach an advanced age.

Age-related use of health care and health expenditure

Most of the national differences in aggregate health care expenditure for older people result from the interplay of general as well as age-related health policy measures, from incentives for providers and for patients, and from regulations, in relation to the treatment of ageing-related diseases (Jacobzone, 1999, p. 2). However, irrespective of those national differences we can observe an age-related increase in the need for health care in all countries.

While health care expenditure – especially on hospital care and medication – begins to rise rapidly at around age 60 (and the average health care expenditure for men often starts to exceed that for women), the expenditure is highest for persons of advanced old age (Figure).

These patterns emphasise why health care is a public policy area of great relevance and economic concern for older people. In the 19th century the income loss on account of health problems dominated the economic consequences of ill-health, which only affected the working population. Today this type of loss is a concern only for employed younger people and for middle-aged people. In case of a severe illness or accident, health care costs even for them can be at least as economically threatening as a temporary income loss. However, unlike the Medicare system in the US, which covers certain health care expenditures only for older people, the UK and Germany provide identical health care goods

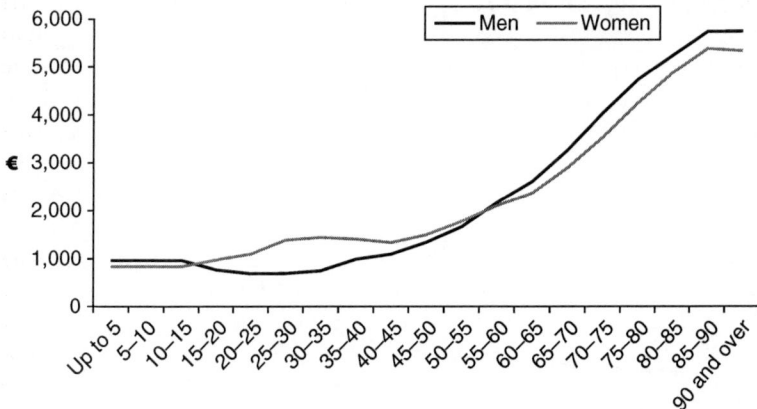

Source: Based on data of the Bundesversicherungsamt for risk adjustment calculations.

Figure 8.2 Average health care expenditure of Public Health Insurance by age and gender, Germany, 2003

and services for the working-age population as well as for retirees via identical institutions and, by and large, according to the same principles. However, differences in need as well as in the ability to pay distinguish the older from the younger groups.

Age-related expenditure patterns may differ somewhat between countries, depending on standards and levels as well as on prices of goods and services. Nonetheless, general differences of this kind can be observed in all countries to a greater or lesser degree. The steepness of the expenditure curves may, however, be affected by the price structure of health services (e.g., fee-for-service as opposed to capitation payments to providers) and direct or indirect age-related rationing. Seshamani and Gray (2002) show that the UK for a long time managed to curb costs especially by *de facto* cutting back on expenditure for older people.

Not surprisingly, the high need for health services, both in the UK and in Germany, makes older people a major target group for health care policy. Looking at the need for services from a social or psychological point of view (rather than from an economic or medical perspective), health care is an area to which the old, more than younger persons, tend to pay more attention and which they consider to be more important. Good health and the ready accessibility of adequate health care is an important aspect of an older person's subjectively perceived quality of life (Walker, 2005).

For many years the average duration of hospital stay has declined in both countries; the average in Germany is, however, still much higher than in the UK. Furthermore, in the past a number of treatments which required inpatient hospital care can today be performed as outpatient hospital care or by the ambulatory health care sector. These shifts can be observed in both countries. However, shorter hospital stays and outpatient treatments lead to a growing need for care – and for household-related social services supplementing medical treatment. As the need for hospital inpatient treatment is controlled more rigorously, patients who no longer need continuous medical attention, but who are still in need of care, may need professional social support when they are unable to call on the informal help of relatives or friends. In the UK this has led to a somewhat stronger accountability of local social services with regard to providing temporary care facilities, in order to reduce the length of hospital stay. In Germany hospitals' social services help to organise temporary care or some ambulatory help. However, the recent hospital remuneration reforms, introducing diagnosis-related groups (DRGs) whose implementation is still on the way, will put further pressure on hospitals to reduce unnecessary 'bed blocking'.

In both countries the major expenditure share is caused by the hospital stay, and in all industrialied countries frequency and length of stay increase with age. Therefore, on the one hand, an increase in the prevalence and the incidence of diseases which require hospital treatment critically affects the need for hospital care and hospital infrastructure – both in terms of the number of beds and surgical facilities available and the investments in new medical technology and know-how. On the other hand, the innovations in treatment methods, for example in anaesthesia, allow established and newly developed medical procedures to be carried out up to a more advanced age than even a few years ago. Both factors contribute to a more-than-average increase in per capita hospital costs among older people.

So far Germany has had an excess of hospital beds in many areas, which has caused high costs, and practically no waiting-lists. In fact, it possibly led to unneeded provider-induced surgical and other procedures in some cases. The policy measures of recent years have aimed at a continuous reduction of hospital beds, or at the closing and the mergers of hospitals in response to changing forms of treatment. The process is bound to continue. In some areas characterised by shrinking and, as a consequence, strongly ageing populations, this may lead to problems, when smaller hospitals in rural areas have to be closed, and the next nearest hospital becomes more difficult to reach.

In the UK *The NHS Plan* should lead to the opposite development. This will be all the more needed because of the rising demand for hospital care on account of population ageing.

Demographic patterns and the state of health

The UK and Germany are 'double ageing' societies – increasing life expectancy is accompanied by a total fertility rate of less than 2.1 – insufficient to maintain the cohort size from generation to generation. However, this process is more accelerated in Germany than in the UK, since Germany has had extremely low fertility rates for almost four decades. While on average a German woman gave birth to 1.34 children in 2003, the total fertility rate in the UK was almost 1.8. This implies that without net immigration both countries will not maintain the generation size (Table 8.1). In the ten-year period from 1993 to 2003 a slight decline could be observed in the UK, while a minimal increase occurred in Germany. In the future net immigration is not expected to offset this natural decline in cohort size.

Life expectancy at birth and at other ages is very similar in both countries and has been increasing over a long time, and quite markedly over

Table 8.1   Demographic indicators, 1993–2003

Gender/age	UK		Germany	
	1993	2003	1993	2003
Male life expectancy at birth	73.5	76.2	72.7	75.7
Male life expectancy at age 60	17.8	0	17.9	19.9
Male life expectancy at age 80	6.4	0	6.3	7.2
Female life expectancy at birth	78.8	80.7	79.2	81.4
Female life expectancy at age 60	21.9	0	22.3	23.9
Female life expectancy at age 80	8.2	0	7.8	8.5
Male prematurely lost life years per 100,000 persons	5,351.11	0	6,271.39	0
Female prematurely lost life years per 100,000 persons	3,272.86	0	3,207.58	0
Total fertility rate	1.8	1.71	1.3	1.34

Source: OECD (2006).

the last few years (Table 8.1). This process is expected to continue in the next few decades. Very likely life expectancy will increase by at least one year per decade. In both countries the life expectancy of women is approximately five years higher than that of men. Life expectancy at birth, at ages 60 and 80 increased in both countries, for men as well as for women and at all ages reported. Differences between the countries are minor.

The number of prematurely lost years per 100,000 men was higher in Germany in 1993, but had become almost equal ten years later. Women, on the other hand, had only slightly fewer prematurely lost years per 100,000 in 1993 in Germany; but the decline in the number of lost years has been more pronounced in Germany since then.

The major effect of differences in fertility rates and the similar increase of life expectancy have led to stronger double-ageing effects in Germany. These differences in the two populations' age composition have consequences for the future demand for health care. Since both countries, like most of the industrialised countries, will be faced by an increasing number and share of older people, combined with increasing health care needs, especially on the part of persons of advanced old age, this development will pose a major social and health policy challenge. However, demographic demand-pull effects will be accompanied by technological supply-push effects, calling for an increased and changing health infrastructure.

Increasingly life expectancy alone is not considered to be an indicator of an improved health outcome. Both for health care and for long-term care we have to distinguish between total life expectancy and healthy life expectancy. In addition disability-free life expectancy can be defined. For UK males, for example, the life expectancy at birth was 75.7 years, the healthy life expectancy 66.8 and the disability-free life expectancy 60.3 in 2001 (Office for National Statistics, no date). (In comparing healthy life expectancy figures of different institutions, one has to keep in mind that different institutions use somewhat different definitions.) Except for women's healthy life expectancy at birth the UK–Germany differences in healthy life expectancy are less than one year (Table 8.2).

The lost healthy years are a little higher in the UK than in Germany. Women lose more healthy years than men in both countries. The difference between total life expectancy and healthy life expectancy indicates that individuals have to expect to require the rather intensive use of health care goods and services for a number of years, and that the associated health care expenditure will be high. The loss in healthy life years is somewhat higher in the UK than in Germany. This may be an indication of slightly better, more generous health care for older people.

The population ageing process has been accompanied by a changing gender composition. As a consequence of both World Wars and their demographic echo-effects, today older males are under-represented: many of their cohorts died prematurely during World War II or thereafter on account of health impairments suffered as soldiers. But both countries' populations are slowly outgrowing this effect. This implies that slowly the number of men of advanced age as well as the number of aged couples will increase, whereas the number of single old women

Table 8.2 Life expectancy[a] and WHO estimates of healthy life expectancy[b], 2002

Gender/age	UK			Germany		
	Life expectancy[a]	Healthy life expectancy[b]	Lost healthy years[b]	Life expectancy[a]	Healthy life expectancy[b]	Lost healthy years[b]
Male at birth	75.4	69.1	6.7	75.9	69.6	5.9
Male at age 60	19.9	15.7	–	19.9	15.9	–
Female at birth	81.2	72.1	8.4	80.5	74.0	7.6
Female at age 60	23.9	18.1	–	23.2	19.0	–

Sources: [a] OECD, [b] WHO.

will increase less rapidly. This will, however, affect the younger old more than those of advanced old age. The latter will continue to be predominantly widowed women. The relative and absolute increase in the number of men of advanced age will, by and large, lead to an increase in medical needs, but less so in social service needs. In most cases they will have a (younger) partner to care for them. On the other hand, women will be more likely to live to an advanced old age, with high need for hospital, other medical and home care needs. Thus the health care and associated social service needs of the aged will continue to have a gender bias.

Both countries will also experience a change in the ethnic composition of the older population. In the UK past immigration patterns are gradually being reflected in the increasing share of the aged with a minority ethnic background. In Germany, 'persons with a migration background' represent a sizeable share of the population as well: approximately one fifth of the population in Germany has such a background (Statistisches Bundesamt, 2006b, p. 74). There may be differences in the morbidity patterns and health resource needs of some minority/immigrant groups. In both countries it is likely, however, that these immigrants will continue to live in their new home countries. Especially in Germany this behaviour pattern was not foreseen originally, when some of the now retired persons of foreign origin were recruited as 'guest workers' – with the naïve expectation that they would eventually return to their country of origin following their retirement. With an increasing number of aged immigrants – sometimes still lacking language skills – hospital care and long-term care will also have to cater for added communication problems between providers and patients.

Ageing and changing patterns in the need for health care

Irrespective of the major causes of an increase in average health care needs and health care costs with age, we observe different types of health problems for people in different age groups. As indicated in the introduction to this chapter, increased population ageing will result in changing patterns of morbidity and as a consequence in changing health care needs. The two countries articulate similar health care objectives – access by 'everybody' to health care (and social) services according to *individual* need – not according to the *individual's* ability to pay. From these general principles it does not follow that the *societies'* or *countries'* needs and their ability to pay have to automatically be equal. Ultimately it is an explicit or implicit policy decision of a country as to which health care priorities and objectives are set, what level of economic and social resources are to be applied to reaching those goals, and who is going to bear what share of the costs of the required resources, i.e., whether and what (re)distribution is needed to achieve these ends.

Major health objectives

The UK and Germany have not been equally ready to formulate specific health policy objectives. Based on the World Health Organisation (WHO) (1998) objectives of 'Health for All', the UK proceeded to formulate health objectives much earlier. In 1997, however, the formulation of health care objectives was partly devolved to the constituent countries of the UK.

The Department of Health's Public Service Agreements and the Department for Work and Pensions' 'Opportunity for All' strategy have, for example, stated the following objectives for England:

- by 2010, increase life expectancy at birth in England to 78.6 years for men and to 82.5 years for women;
- substantially reduce mortality rates by 2010, including:

  o from heart disease and stroke and related diseases by at least 40% in people under 75;
  o from cancer by at least 20% in people under 75;
  o from suicide and undetermined injury by at least 20%;

- reduce health inequalities by 10% by 2010 as measured by infant mortality and life expectancy at birth;
- reduce adult smoking rates to 21% or less by 2010;
- halt the year-on-year rise in obesity among children under 11 by 2010;
- reduce the under-18 conception rate by 50% by 2010 (*United Kingdom National Report for the Open Method of Co-ordination of Health Care and Long-Term Care*, 2005, p. 8).

These objectives do not specifically address the very old. In fact, some goals refer to persons up to age 75 only. There is a stronger emphasis on the young.

The Department of Health (2006) stated further objectives in the context of *The NHS Plan* which include various aspects of performance, such as maximum waiting-times and greater choice for patients.

With regard to older people we can find specific objectives in the National Service Framework for Older People (Department of Health, 2001b). The first objective states that older people should not be denied services on account of their advanced age. Given previous practices this should mark a turning point in UK health policy in favour of this group.

Germany has been rather slow and hesitant with regard to specifying health objectives, possibly because responsibilities and competencies are fragmented in a federal state (Wildner and Weitkunat, 1998).

Only Public Health Insurance is based on very general principles similar to those of the UK: the necessary services should be provided on the basis of need and financed according to the insured member's ability to pay (SGB V, §§ 2, 3 and 12). A number of specific measures convey implicit objectives. With regard to older people the general principles imply that they should be provided with health care according to their greater need and that they should have to pay less than younger insured members on account of their lower income.

Only recently has an effort been made to develop more concrete health objectives (SVRKAiG, 2002; Rosenbrock and Gerlinger, 2006, p. 52). An ongoing project, 'Gesundheitsziele.de', aims to develop national health objectives.

At this stage the following subject areas have been agreed upon:

- diabetes type II: reducing risks, early diagnosis and treatment;
- breast cancer: reducing mortality and improving quality of life;
- depression: prevention, early diagnosis and treatment;
- reducing tobacco consumption;
- growing up healthy, nutrition, exercise and coping with stress;
- empowerment and patients' rights (Gesundheitsziele.de, 2006).

While the formulation of national objectives does not explicitly address older people as a target population, diabetes type II and depression certainly are diseases with a high prevalence among older people. Individual provinces (*Länder*) of Germany have formulated their own provincial objectives, relating to the *Länder*'s vs the national responsibilities.

Alternative instruments to curb excessive aggregate resource use

Health care expenditure constitutes one indicator of the generosity and accessibility of a health care system. A strong positive correlation between per capita income or GDP and health care expenditure or the share of health care expenditure in GDP can be observed. Both the UK and Germany are certainly wealthy countries. But even the past somewhat higher per capita income or GDP in Germany cannot fully explain the different levels of health care expenditure of the two countries. They also result from the fact that the UK used to be rather thrifty and contain health care costs, i.e., her total health care expenditure and the expenditure's share of GDP were much lower than in Germany. Viewed differently, the UK system worked more economically, while the German system tended to use resources more generously – less efficiently or less cost-effectively. The latter interpretation is also reflected in the WHO ranking, according to which the inter-country differences between the UK and Germany are not high (WHO, 2000, pp. 152–5). For example, performance measured in terms of mortality was almost identical in the two countries despite the different levels of expenditure on health resources (Table 8.3).

This results from the rather strict expenditure control employed in the UK. For example, the number of hospital beds and the number of physicians per 1,000 persons was still much lower in the UK than in Germany in 2004 – after some years of more rapid growth of expenditure in the UK than in Germany. As outlined above, however, the UK has set out to increase its health care capacities and expenditure considerably under *The NHS Plan*, which should show effects during the next few years.

All countries have to apply instruments to ensure efficiency and to curb costs in the face of increasing needs and options for health care.

Instruments to contain the demand for health care – co-payments

The German health care system encompasses strong incentives to providers towards an expansion of services. Even the imposition of

Table 8.3   Health care expenditure and health care resources, 2004

Item	UK	Germany
Health expenditure as % of GDP	8.1	10.6
Health expenditure per capita in $ (PPP)	2,508	3,043
Acute hospital beds per 1,000 persons	3.6	6.4
Physicians per 1,000 persons	2.3	3.4
Practising nurses per 1,000 persons	9.2	9.7
Hospital staff per bed	6.50	2.03
Nurses per bed	1.80	0.75
CT scanners per 1,000,000 persons	7.0	15.4
MRI scanners per 1,000,000 persons	5.0	6.6

Source: OECD (2006).

sectoral budgets on providers has proved insufficient to curb expenditure effectively. Moreover, it cannot be denied that the parallel coverage in the public and the private health insurance leads to competition between the two systems. This tends to raise the level of benefits as well as associated expenditure and does not lower prices or costs. Moreover, the low (or zero) direct costs at the point of delivery eliminate the price-incentive mechanisms, leading to moral hazard, on the part both of the insured and the providers.

If moral hazard is a logical consequence of insurance coverage, imposing partial 'price components' in the form of co-payments may curb excess demand for and expenditure on health care. So far Germany has followed a mixed cost containment policy: it includes provider-oriented incentives and budgeting, but also attempts to reduce demand, e.g., by increasing co-payments by patients. This process has led to a gradual increase, starting from 1981 and culminating in the measures introduced by the Public Health Care Modernisation Act of 2003. Until 2003 co-payments had been combined with relatively generous exemptions for low-income groups and the chronically ill. (No co-payments are applicable to children.) Even though there was no formal exemption for older people, in effect the income and morbidity criteria benefited many persons past retirement age.

In 2004 a significant increase in co-payments as well as a reduction in income-related exemptions took effect. This involved an increase in co-payments for the first four weeks of hospitalisation, a (modest) quarterly fee for physicians' services, an increase in co-payments for medication and some further minor co-payments. The maximum co-payments were confined to 2 per cent of income (1 per cent for the chronically ill), to be borne by all adults. The elimination of the total income-related exemptions led to a marked increase in direct payments by the low-income old (Pfaff *et al.*, 2003). Despite this increase one has to admit that even after these measures took effect, the co-payment level is comparatively modest by international standards.

Apart from prescription charges the UK has hardly any co-payments. For the most part the NHS philosophy stands for provision of services without direct payment at thepoint of delivery. However, in recent years dental services have been privatised to a large extent; an estimated one half of the costs have to be borne by patients (Busse *et al.*, 2005, pp. 40–1). Considering the age-specific patterns of demand, this affects persons of middle age more than older people.

Instruments to reduce capacity – waiting-lists and rationing

The NHS allows more centralised, direct control of capacities and expenditure, i.e., control of the supply rather than of the demand for health care. The control of available resources – by making them scarce – was one way to curb health care expenditure. Since the objectives of the NHS neither desired nor allowed an allocation process via the price mechanism, alternative means of 'rationing' goods and services had to be applied. The objective of providing services according to need implies that more needy persons should be treated before the less needy. While in principle this rule of action is convincing, it is difficult to implement in practice: what, after all, determines need? Is an older person less needy than a younger one, a home carer more than a single person, a member of the labour force more than a retiree, an otherwise healthier person more than a person in general poor health? Equity rules that satisfy social justice are not unequivocal.

When no immediate mortal danger arises from a delay of treatment, one rule satisfying social justice would be the 'first come, first served' principle. As long as capacities are adequate in terms of quantity and quality, the necessary waiting-time would not be excessive. However, the growing range of diagnostic and therapeutic measures would gradually result in a growing need for more and more modern equipment and for sufficiently well-trained personnel. If one adds to that a gradual ageing of the population, a further demand-pull effect leads to growing shortages. Therefore the shortages of health care resources have resulted in considerable and growing waiting-times for elective procedures – especially for older people. This has led to considerable dissatisfaction on the part of the population and has become a major annoyance in the UK, sufficiently so that recently some patients have even been treated abroad at the expense of the NHS in order to reduce waiting-times.

A waiting-list problem does not really exist in Germany. There may be some minor waits for doctors' appointments. In this context some policy makers as well as some scientists have criticised the fact that many doctors grant appointments to privately insured patients much faster than to patients covered by Public Health Insurance. First of all, they can charge higher fees for the privately insured, and secondly, they are not in danger of exceeding a desirable volume of services agreed upon between the physicians' association and the Public Health Insurance funds during a given quarter or year. In fact some hospitals in Germany treat foreign patients because excess capacities are available. In the border areas with the Netherlands, for example, a chain of 13 hospitals has signed a contract with the Dutch health care fund Amicon for the treatment of Dutch patients.

Shipping patients abroad, in order to shorten waiting-lists for patients in urgent need of treatment, is one way to deal with the problem. Giving preferential treatment to some patients, for example because of their private insurance and their higher ability and willingness to pay, is another. The latter could, in fact, be of benefit, provided the additional financing led to an expansion of the available resources and shortening of waiting-lists (Hoel and Saether, 2003). This could be the case in a health care system like the UK's. In the German system this has not been the case. On the contrary, given fixed capacities, the preferential treatment of private patients will tend to lead to longer waiting-times for those in the Pubic Health Insurance system. Since in the German system, for historic reasons, a larger share of the older than of the total population is covered by Public Health Insurance, this type of preferential treatment of the privately insured will result in disadvantageous effects for older people.

Waiting-lists are certainly annoying even for elective procedures. They may also lead to less advantageous health outcomes. However, some studies indicate that these do not significantly increase mortality. Moreover, a strong relationship between capacity and the length of waiting-times could be observed (Hurst and Siciliani, 2003; Siciliani and Hurst, 2003).

The NHS has responded to the problem of excessive waiting-lists by increasing capacity, on the one hand, and by providing more transparency, on the other. Moreover, people who have had to wait for six months are offered a choice of provider, in order to reduce further waiting.

Whenever a resource is scarce some process of selecting those who are to get the benefit of the service, i.e., some form of rationing, has to be applied. The 'first come, first served' rule, determining the position on a waiting-list, is but one applicable allocation rule. It certainly has the advantage of satisfying one principle of justice, i.e., equality of treatment for all. Willingness and ability to pay is not considered to be a 'fair' or acceptable criterion in all those countries that have opted for a predominantly publicly financed health care system. Even partial pricing via co-insurance does not allow for the full operation of this principle. In the context of health care both the UK and Germany do not consider it as a desirable primary allocation criterion.

Using some indicator of need as a criterion poses an alternative. In case of emergency treatment an immediate need is evident. In case of other treatments the allocation is less evident. With the advance of medical knowledge a wide range of services became available, allowing the treatment of an increasing number of disorders for more and more people. This raised the question whether every possible diagnosis and therapy should be made available to every person, possibly even when a rather minor subjective benefit can be conveyed only at high social cost.

In the context of this chapter, the interesting question is whether rationing according to age limits is one 'fair' or acceptable criterion for rationing or for denying benefits. The initial contribution by Callahan (1987, 1996), suggesting that beyond a certain age certain benefits should not be granted, has led to a heated discussion. In practice age-related rationing has occurred both in the UK and in Germany. In the UK it was a more open form of managing scarcity. However, the UK's National Service Framework for Older People stated very clearly that in future services should not be withheld on account of a patient's age (Department of Health, 2001b).

In Germany, where scarcity was less pronounced, age rationing took place in a more indirect way. Very often the higher risks associated with co-morbidity led physicians to deny certain treatments to aged persons. On the other hand, the opposite could equally be observed: older persons were sometimes subjected to strenuous and ultimately not very useful treatments – sometimes to justify a hospital stay. However, with regard to some treatments for which temporary shortages occurred, e.g., severe trauma treatment, decision makers did often give preference to the treatment of younger persons, because the prospects of success in terms of health outcome were higher.

The discussion about age-related rationing ties in very closely with age discrimination. It is not only an issue of economy vs ethics or equity (Breyer and Kliemt, 1994). Age-related rationing in Germany was also discussed with regard to the general benefit basket which the Public Health Insurance funds offer. Advocates of rationing argue that in the future, for currently younger persons, one could reduce the general benefit basket beyond a certain age and encourage the cohorts affected by this cut to take supplementary insurance to cover the more extended benefit basket in advanced age (Breyer and Schultheiss, 2002).

The increased efforts to introduce quality control and standards may have an indirect impact on negative or positive age discrimination as well. So far certain diagnostic measures, e.g., cancer screening, are provided free of charge beyond a certain age or in case of an increased risk only. Some of the guidelines based on cost-benefit analysis, for example of drugs, may also lead to different guidelines with regard to younger people and the elderly. Given the different prevalence and prospects of treatment of diseases in different age groups, guidelines of the UK's National Institute for Clinical Excellence (NICE) may well recommend a given good or service for one group but not for the other. Given the higher prevalence of most diseases in old age, it is more likely that the application of such a rule will lead to positive discrimination in favour of older people. However, the outcome of the underlying cost-benefit consideration is crucially dependent on how costs and benefits are defined. In that sense quality considerations could also be turned into an instrument of negative age discrimination.

So far Germany has no equivalent institution to NICE. The Institut für Qualität und Wirtschaftlichkeit im Gesundheitswesen (IQWIG), which was established in 2004, was at first not allowed to carry out cost-benefit analyses. Only the latest reform of 2007 added this task to its previous competencies.

The differences in health policy and their effect on older people

Historical developments and political preferences account for different health care systems in the UK and Germany. In spite of very similar general objectives, the two countries have chosen different policy instruments and different levels of generosity of provision. But in the context of the major reforms of the last decade we note a convergence of principles, instruments and expenditure levels, while both countries continue to hold on to their systems in general. Specifically the following aspects warrant a commentary:

- accessibility of health care;
- the role of primary vs secondary care;
- managing the tradeoff between economy and quality;
- the differential treatment of the aged and of younger persons;
- the trend towards privatisation of health care;
- transparency, patients' choice and patients' rights.

Accessibility of health care

Accessibility may be defined with respect to a person's entitlement: who has access to health care? Alternatively, a more concrete definition may be used: does everyone *actually* have access to the same benefit basket? The UK and Germany emphasise different aspects.

Since its establishment in 1948 the NHS has always been a universal system in terms of encompassing the entire population within the same institutional framework. Germany only passed a law requiring *every* resident to have health insurance as late as February 2007. Even thereafter two separate systems will continue to coexist, which are characterised by different principles: public and private health insurance for the most part apply to different groups. Both the groups, however, include all age ranges. One of the Public Health Insurance system's main advantages for older people is the socially desired redistribution to them; approximately half of the health expenditure for this group is borne by younger insured persons. Private health insurance does not entail an interpersonal redistribution of this type; it is in principle set up as a system which does not intend an interpersonal redistribution *ex ante*. Of course, viewed *ex post* there is always a redistribution from healthier persons to sick persons.

In order to contain health care costs to a desired level, in the past the UK system managed the *actual* access to health goods and services for all residents on the basis of rather strict budgeting and capacity restrictions. For a number of years the NHS did not allow provision to rise in line with actually growing, more generously defined needs. For all practical purposes the allocation of scarce health resources led to a certain degree of age discrimination. Some treatments were denied to people beyond a certain (advanced) age, and they were subjected to long waiting-lists. This policy is reflected in the less than proportionate increase in health care expenditure for older people in the past (Goddard and Smith, 2001; Seshamani and Gray, 2002).

Moreover, the rather limited provision entailed a disadvantage to the larger share of the older people with lower incomes who could not afford additional private insurance or high out-of-pocket payments. The incidence and prevalence of some diseases which require timely treatment on account of a continuous progression of the disease (e.g., cancer), are higher among the older population. Therefore more of them suffer disadvantages on account of long waiting-lists. The recent reforms, which will lead to an increase in capacity, with shortening of waiting-lists and explicit denial of age discrimination, should assuage these negative effects on the aged.

In Germany, the recent reforms requiring health insurance for everyone were combined with the choice of a more reasonable insurance coverage for the privately insured older age group (*Basistarif*). This will extend the accessibility of health care to the entire population. The discussion about the possibility of not granting every possible benefit to persons of all age groups and all health states has only come up rather recently. The issue has been discussed not only from an economic point of view, but also with regard to humanitarian considerations. For the most part this discussion is not motivated by the desire to deny benefits *per se*, but by considerations of additional financial resource requirements, either in terms of higher contribution rates for retirees or by requiring additional, voluntary insurance. Ultimately recent discussions about reforming the financing of Public Health Insurance include the attempt to justify higher contributions by older people. This is true for the reform suggested by the Christian Democrats (CDU and CSU, 2004) to change to lump-sum contributions (*Kopfpauschale, Gesundheitsprämie*) as well as for the reform suggested by the Social Democratic Party (SPD, 2004), the Green Party, the left-leaning parties PDS and WASG and also by the trade unions to introduce compulsory Public Health Insurance for all, combined with levying contributions not only on wage and wage replacement income but also on capital and other income (*Bürgerversicherung*). The reform passed by the Grand Coalition of the CDU/CSU and SDP is a compromise between the two, which, however, did not encompass those parts which would have imposed relatively higher financial burdens on older people.

The UK and Germany converge in the sense that coverage for all will improve in Germany and the same improved benefit basket will actually be available to older people and to younger persons in the UK.

Primary care and secondary care

The UK provides an effective primary care system which includes a gatekeeper and steward function, especially on the part of GPs. Every patient has to register with a GP in order to receive free services. The GP also prescribes medicines and refers his or her patients to specialists or hospitals, where similarly no fees have to be paid at the point of delivery. In case of an emergency the patient can receive treatment without prior referral. This system can in principle operate very efficiently. Similarly patients can register with dentists. However, as we have seen, in recent years the costs of dental services have been privatised to a large extent.

Generally a well-operating system of primary care has the advantage that it is fairly economical. And, in principle, a qualified person guides the patient through the treatment pathway. Generally the remuneration system will have some incentive and disincentive effects, which tend to influence the health process and the health outcome. For the most part doctors are under contract to the NHS or an NHS-funded institution.

The reforms commenced during the last few years have not changed the differentiation between primary and secondary care, including the gate-keeper function, but have modernised the system and separated purchaser (for the most part PCTs) and provider functions (GPs, specialists, hospitals, etc.).

For older people this system has the great advantage that, in case of multi-morbidity and chronic diseases, a well qualified GP can provide expert guidance at a reasonable price. In the past the shortages have, on the one hand, ensured economical provision of services; and they also provided incentives for referrals. On the other hand, GPs were not only able to deal with the more straightforward problems directly, but were also – so to say – the gate keepers of waiting-lists. The separation of purchasing and providing could in principle enable the purchasers to choose suitable providers. This can introduce elements of competition. However, the presence of shortages will not allow the purchasers to select better quality providers.

The differentiation of primary and secondary care, with the primary care gate-keeper function, could in principle provide a suitable framework for coordinating medical and social services. A more integrated approach than what the German system offered was, in fact, characteristic of the UK one. Of course, in practice it often did not work in a satisfactory manner. Therefore the recent reforms tried to eliminate organisational barriers, by devolving the central responsibilities for health care services more to the local level. This placed medical and social services under the control of the same level. There is hope that the decentralisation of responsibilities for the provision of medical services will enable the local agencies to coordinate their efforts better. This may of course lead to greater regional diversity in the level and standard of services. For older people as an important target group of health policy, the improved organisation is of particular relevance, since they are more likely to move between primary and secondary medical services and social services.

Germany is characterised by a somewhat different structure. It also has doctors who work as GPs. But in addition to these, an increasing number and share of the doctors have worked in ambulatory care as specialists. Until the mid 1990s, the insured received a voucher from their Public Health Insurance fund per quarter, which they had to present to the first doctor whom they went to see during the quarter. In case of need, this doctor would then have to give them a referral voucher for other doctors or for hospital or other services (like physiotherapy). Frequently the first doctor to see the patient was a GP. But in principle it was also possible to go to see a specialist directly. The voucher entitled the patient to receive treatment free at the point of delivery. Doctors were not remunerated on a voucher-based capitation fee, but they needed the voucher to be able to receive their fee-for-service remuneration.

Gradually patients got more control over their vouchers. Eventually electronic insurance cards were introduced which allowed patients to see any doctor without referral. This loss of control by one doctor undermined the gate-keeper function – which had not been as developed as in the UK in the first place – and led to an increase in costs and to a lack of coordination in treatment and medication. On the other hand, the element of greater choice and autonomy on the part of patients is considered to work to their advantage.

The ensuing financial and medical problems have led to the introduction of measures to strengthen coordination of services and to move into the direction of primary care. A co-payment (*Praxisgebühr*) of €10 per quarter for doctors' services was introduced in 2004. Persons who got a referral from the first doctor they saw during a quarter had to pay the fee only once per quarter. On the other hand, efforts were made to introduce a primary care system (*Hausarztmodell*) on a trial basis. The reform of 2007 liberalised the rules for contracting between Public Health Insurance and the Association of Doctors Holding Contracts with Public Health Insurance (Kassenärztliche Vereinigung). So far the doctors' association has had a near monopoly position in contracting with Public Health Insurance. In the future more differentiation has to be offered. Each fund will have to offer a primary care tariff to its insured members, on the one hand. On the other hand, special rules will apply for doctors who are GPs or who opt to work as GPs (specialists for general medicine or for internal medicine) in the context of a primary care-type service provision.

A change of the remuneration system, essentially away from the fee-for-service to a differentiated capitation system, is under way. It is too early to foresee how this system will work. Potentially it may be an improvement for older people, provided the GPs are sufficiently well qualified and that the remuneration system does not keep them from necessary referrals to specialists.

Unlike in primary care in the UK, the German system has so far been characterised by the work of self-employed doctors, most of them working in individual private practices. But over the last decade a number of reform steps have been taken to encourage cooperation between doctors of the same specialty and across specialties. This could be of advantage for the chronically ill and for persons with more than one disease – both in terms of convenience and in terms of better quality. In the past German health care delivery suffered from a rather strict separation of the ambulatory and the hospital sectors. Several of the reforms have attempted to break down this 'wall'. There have been some improvements, but the whole system is undergoing a major change in the wake of the introduction of the DRG-remuneration system in hospitals, which is still in transition, and the reform of 2007.

By and large – unlike in the UK – many of the structural reforms in Germany have been motivated by the attempt to cut down capacities and to reduce a 'provider-push' cost effect. While the UK system's expansion of capacities should make it easier to improve service provision, the reduction of capacities in Germany entails potential dangers to the quality of health care there.

The coordination between medical and social services involves problems in both countries. They are potentially greater in Germany, because health care is dominated more by the clinical approach than by a more integrated (public health) approach; particularly for older people it can cause deficiencies in their health service provision, since they frequently require a more integrated service basket, allowing for transitions between hospital, ambulatory care and long-term care.

The increasing prevalence of mental disorders in persons of advanced age poses a great challenge in both countries. Both lack awareness of the problems and adequate services to deal with them. Moreover, some of the disorders are aggravated by social conditions (poverty and isolation in old age).

Managing the tradeoff between economy and quality

The attempt to streamline the health care system and to increase efficiency and economy entails the danger of reducing quality. Adequate quality control is therefore of great relevance.

Both countries have taken steps in this direction – and both need these measures. With regard to hospitals, both internal and external measures of quality control have been introduced, which also involve greater transparency. In the UK the establishment of NHS Foundation Trusts with earned autonomy should provide incentives. The publication of quality control reports has similar effects – provided patients are offered some choice.

The introduction of drugs is subject to a number of strict regulations, both at the national and the European Union level. Furthermore, the two institutions NICE and IQWIG are charged with carrying out cost-benefit analyses especially for drugs. For older people a possibly dangerous development may result, if the publication of performance indicators encourages a risk selection on the part of hospitals and an unwillingness of hospitals to treat high-risk patients. These risks cannot be avoided completely. Therefore attention has to be given to how incentive systems should be designed. The measures providing incentives for economic and efficient operation are no doubt needed. The formalisation of performance indicators may, however, entail negative side-effects. For example, if complication rates are published, a hospital may prefer not to treat those patients for whom complications are more likely to occur. Remuneration systems may also provide incentives to avoid treating older people suffering from various other ailments.

Differential treatment of older and younger people

In the past the differential treatment of older people was sometimes more common, because some diagnostic or therapeutic measures were part of the benefit basket only beyond a certain age. This is of special relevance in achieving an early diagnosis of dangerous, progressive diseases. This form of positive discrimination is observable in both countries and it is likely to continue. The form of quality control exercised by NICE, for example, may lead to guidelines recommending a certain treatment for older people only, because the cost-benefit evaluation of medication may indicate that the treatment is only called for in case of higher risks, which frequently occur in advanced age.

The different pattern of diseases in old age leads to higher risks associated with certain treatments in old age. Two examples will be cited, which are by no means the only ones. As already mentioned above, mental disorders which occur frequently in advanced age have often not been diagnosed. This is particularly true of depressions, which are frequently not recognised as a disease. This shortcoming has been recognised in the UK as well as in Germany. In both countries the health objectives include a priority for dealing with this shortcoming. This involves a multifaceted approach, including better training of medical personnel, a better integration of medical and social services and an adequate support system for carers. The second example of special problems for older people refers to the consequence of multi-morbidity. Very often a considerable number of medicines are prescribed to a member of the older age group, because they simultaneously suffer from different diseases. In this case the side- and interaction effects of these medicines are insufficiently known – both to the pharmaceutical industry and to the individual doctor prescribing them. Possibly a more generous system runs an even higher risk of providing excessive, unbalanced treatments – sometimes making people more sick than with less medication.

Apart from these two examples relating to illness, generally older people can be more seriously affected by a lack of coordination between health and social services, especially when illness and the need for long-term care simultaneously affect individuals.

Trend towards privatisation of health care

Germany has always been characterised by a large sector of private service provision and a private insurance sector which offered full insurance coverage to part of the population. Most likely the long periods of predominantly Conservative governments encouraged this development. More recently, a wave of different forms of privatisation took place in the hospital sector. If anything, the reforms of the last few years encouraged this tendency. The UK tended to view health care more as a public good that needed not only public financing, but also public sector production. However, recently certain steps have been taken to privatise some activities as well.

Transparency, patients' choice and patients' rights

The new media offer means of communicating relevant information that could not be envisaged a few decades ago. In their efforts to modernise their health care systems, the UK and Germany both try to make use of these means in a variety of ways in order to enhance the transparency of the system. These efforts are used to improve quality and to empower patients in exercising their rights and in their ability to participate more actively. Both countries make greater efforts to look at patients more as customers than as recipients of benefits in a paternalistic system. The other side of this coin is that more individual responsibility is also placed on patients. This involves a greater emphasis on prevention efforts on the part of individuals and a greater responsibility in decision making (the model of informed consent). In the German setting, self-reliance was often interpreted only as financial self-reliance – the obligation to pay co-insurance in order to reduce moral hazard.

For older people the greater emphasis on transparency has positive and negative effects. On the one hand, as major users of the health care system they could benefit from better information and improved rights. On the other hand, at this stage older people frequently have not adapted as much as younger persons to the use of instruments for gaining information about health-related matters ('e-health'). This is especially true of the less educated. The advance in information technology may therefore turn out to be an added instrument for increasing social inequality. Apart from the expertise to use modern information-gathering methods, the specific morbidity of persons of advanced age also indicates a higher prevalence of cognitive impairments. Therefore even educated older persons may not be able to avail themselves of new technologies as efficiently. Since information and transparency are major instruments for exercising patients' choice and patients' rights, it may be more difficult for the aged to avail themselves of these instruments.

In the UK, considerable efforts are directed at communicating information about waiting-lists and an increased choice of providers. These are aspects of immediate importance to many patients. In the eyes of the average patient this should certainly lead to a perceived quality improvement. In Germany, this aspect is less relevant. However, information about quality will play a more important role – at this stage for hospitals, in the future probably more and more for ambulatory care as well. For patients the availability of information is an important prerequisite for exercising choice in a sensible way.

For older people in the two countries there seems to be a difference in emphasis about the use of information and of greater transparency in order to improve the opportunities for timely diagnosis and high-quality treatment. In the immediate future, the UK will be faced more with the problems of making services available and improving logistics. In Germany, the efficient exercise of choice and avoidance of discrimination in potentially emerging processes of risk selection will be concerns of the future.

Conclusion

The differences between the UK's and Germany's health care systems have some impact on accessibility and the quality of outcome for all patients. For older people some of these features have a greater impact, because they are major users and they show somewhat different morbidity patterns. The organisational differences point to the fact that the UK system has been better coordinated (between sectors and between health and social services), with the UK showing more integration of various components of the services (on account of a unified NHS) and, by design, the UK espoused a higher degree of equity (equal access for everybody) than in Germany. These advantages of the UK system, however, were outweighed by the fact that the system became increasingly underfinanced – possibly as a consequence of the political control of a more monolithic tax-financed system. The German system showed a number of inefficiencies and intrinsically some lack of equity. This was, however, compensated for to some extent by a number of positive features: by greater elements of competition, by more choice and by more generous provision. For older people, the advantage of higher equity in the UK system was offset by the shortages resulting from age discrimination. The German system suffered from its emphasis on the treatment of acute diseases and from inadequacies in prevention and the treatment of chronic diseases.

Both countries have undergone major reforms during recent years which, in effect, indicate a convergence of the two systems, while maintaining major differences in their general design. It is likely that both systems will become more efficient in the provision of health care and also more geared to the needs of older people. Some known problems have been tackled. Nonetheless, both societies will be faced by economic as well as ethical constraints in the future. It is not quite evident yet what tradeoff they will choose with regard to the older age group.

9
Long-term Care in Germany and the UK

Caroline Glendinning and Gerhard Igl

Context – current and future demand in the UK and Germany

Current figures and future demographic projections in the UK and Germany are quite similar. By 2030 around a quarter of the population of each country is expected to be aged 65 and over and one in 16 people will be 80-plus (Table 9.1).

Future demand for long-term care in both countries is therefore expected to increase substantially over the next three decades. In the UK, between 2001 and 2031, the numbers of residential and nursing-home places in England are estimated to increase by 58 per cent and the number of home care service hours by 57 per cent, simply to keep pace with demographic pressures (Comas-Herrera *et al.*, 2004). In Germany, a 51 per cent increase in the numbers of very old people (80-plus) is expected between 2030 and 2050, even though the total number aged over 65 will remain relatively constant (Rothgang, 2003). Projections of the impact of these demographic trends on the German long-term care insurance scheme suggest that total expenditure is expected to increase by around a third between 1995 and 2030 (Schmähl and Rothgang, 1996).

Scope of long-term care

Long-term care in both countries covers assistance with personal care – dressing, bathing, preparing meals and cleaning; nursing; and rehabilitation and therapy services. In Germany, long-term care also includes help with domestic and house-keeping tasks. In both countries, long-term care is mainly provided in older people's own homes

Table 9.1 Older people in the population, 2000–2030

Country	2000	2000	2020	2020	2030	2030
	% age 65+	% age 80+	% age 65+	% age 80+	% age 65+	% age 80+
Germany	16.4	3.6	21.6	6.3	26.1	6.8
UK	16.0	4.2	19.8	5.1	23.1	6.5

Source: Karlsson *et al.* (2004), derived from UN data.

or in residential and nursing homes; day care and hospital care are less common.

In both countries, the greatest volume of long-term care is provided by close relatives (Pickard *et al.*, 2000). Increasing proportions of informal carers of older people in both countries are themselves elderly, as male life expectancy improves and care-giving by spouses becomes more widespread. Almost 16 per cent of people in England and Wales aged 50 and over provide unpaid care for family members, friends or neighbours. One in four of these spend 50 hours or more a week caring, while 50 per cent of carers aged 85-plus spend at least 50 hours a week caring (Office for National Statistics, 2004). In Germany, more than 70 per cent of all 'care-dependent' people and 66 per cent of older 'care-dependent' people living at home rely wholly on informal care (Rothgang, 2003); indeed Schneekloth and Müller (2000) found that only 4 per cent of dependent people in private households did not have at least one informal carer.

Funding and organisation of long-term care in Germany

Pressures for change

Debate about the introduction of long-term care insurance (LTCI) in Germany stemmed from the 1980s, when increasing numbers of older disabled people began to place severe pressures on both sickness insurance and the social assistance scheme funded and administered by regional *Länder* and local municipalities, that covers the costs of care for poorer older people.

All German employees are required to belong to a sickness insurance fund. However these funds operate a very rigid distinction between medical treatment and 'care' and do not cover the latter. Until LTCI was introduced, funding for long-term care was partial and fragmented:

• Sickness insurance covered the costs of medical, nursing and personal care, if prescribed by a doctor, for up to four weeks after hospital treatment, especially if this facilitated prompt hospital discharge. This created a 'revolving door' where older people were readmitted to hospital in order to re-qualify for four weeks' funding for home-based care.

- Regional and local social assistance schemes covered the costs of care at home or (more commonly) in an institution. The level of benefit depended both on levels of resources (taking into account the financial and in-kind contributions of other family members under the subsidiarity principle) and on the level of help needed. Assessments were carried out by medical officers on behalf of the social assistance offices; there were no nationwide rules governing the assessment. In 1990, 80 per cent of people in institutions in the former FRG and 100 per cent in the former GDR relied on social assistance for the costs of their care; in turn, this constituted one-third of total expenditure on social assistance in the former FRG.

- Some *Länder* had introduced locally funded additional benefits (*Landespflegegeld*) to cover long-term care. In Berlin, for example, the level of the benefit was not means-tested, but depended on the assessed level of 'helplessness' or 'destitution'; from 1985, beneficiaries were also entitled to additional 'assignments' of domiciliary services (intended to provide a break for family carers).

The restricted availability of funding for long-term care, together with the strong principle of subsidiarity that placed legal responsibilities on households for the care and support of family members, meant that services were extremely limited, in both levels and variety (Schunk, 1998):

- Weak entitlements to funding for home care services meant these were scarce and piecemeal; the main type of service was institutional care.
- Mechanisms for planning and coordinating local services were weak.
- The main service provider organisations effectively operated a 'cartel', dominated by organisational interests rather than users' needs.
- Health insurance covered only a fraction of the amount of care needed by many older people. If they needed home care services or institutional care (for example, because informal support was not available), they had no alternative but to purchase these privately and 'spend down' their assets in order to qualify for social assistance. The substantial role of social assistance was a major financial burden

on the *Länder* and municipal governments and was also considered highly stigmatising.

- There was no social protection for the relatives who provided virtually all non-institutional care.

This piecemeal provision generated pressures on health insurance (which was itself experiencing major problems in controlling costs and managing human resources) and on municipal and *Länder* social assistance budgets. The latter could not be controlled locally, as federal laws determined their eligibility criteria and coverage. Moreover, within the context of the German welfare state, spending down to become eligible for social assistance was regarded as incompatible with basic citizenship rights. Debate therefore focused on how universal coverage could be achieved and this became the key feature of German LTCI.

Raising revenue for long-term care

Until 1994 there were four social insurance schemes covering pensions (old-age, survivors' and invalidity), sickness, unemployment and industrial accidents. A fifth scheme covering long-term care was phased in from 1994. Although LTCI is separate from health insurance, it is administered under the umbrella of the sickness insurance funds.

LTCI is financed by adding a further 1.7 per cent of gross monthly payroll (up to an income ceiling of €3,600 per month [2008]) to existing social insurance contributions. Social insurance contributions are normally divided equally between employers and employees. However, because of the staunch resistance of employers to additional payroll costs, the employers' share was compensated for by the abolition of one day's statutory paid holiday, equivalent to about 75 per cent of the employers' contribution. However, the LTCI premium is small compared with other social insurance premiums and it cannot be increased without federal legislation.

Because membership of LTCI, like health insurance, is mandatory, almost the entire population is covered; non-employed family members are covered by the head of household's contributions. Indeed, its universality is the key feature of LTCI, which is popularly known as *Volksversicherung* ('people's insurance'). The contributions of unemployed people are paid by their unemployment insurance. Until 2003, half the care insurance contributions of retirement pensioners were paid by their pension insurance and pensioners themselves paid the other half; from 2004 retired people were required to pay the full 1.7 per cent contribution themselves. From 2005 persons without children had to pay an additional 0.25 per cent contribution. This was due to a decision of the Federal Constitutional Court which found an unequal treatment between persons with and without children. Around 10 per cent of employed people belong to private care insurance schemes; however, these are legally required to offer coverage, contributions and benefits on equivalent terms to the statutory LTCI scheme. About 2 per cent of the population belong to insurance schemes for specific occupational groups.

Organisational arrangements for long-term care

Federal legislation regulates the insurance funds that are jointly administered by employers and unions. The levels of LTCI contributions, the global budget for LTCI benefits and the levels of benefits are all fixed by federal law, regardless of fluctuations in inflation or wages. Federal government also controls the 'care dependency' entitlement guidelines, thus giving it substantive regulatory and cost-controlling powers that do not apply to other branches of social insurance.

The LTCI funds are independent 'departments' within the sickness insurance organisations. They are responsible for collecting members' contributions, determining members' eligibility for services and reimbursing providers for home and institutional care, within the limits of their respective budgets.

Länder-level associations of LTCI funds (and municipalities, as payers of social assistance) negotiate annually with associations of the organisations that provide professional home care and institutional care services, over service contracts and prices. Prices exclude building and infrastructure costs, which are the responsibility of the *Länder*. However, this responsibility is discretionary, so funding for capital developments, care management services and other elements of service infrastructure is at substantial risk of territorial inequities. Moreover, because nursing homes are normally private enterprises, *Länder* subsidies are restricted by EU law. In the annual price negotiations, federal law requires that priority be given to private and charitable provider organisations over public providers, in order to stimulate market-style competition. Because the role of local municipalities in funding long-term care through social assistance has diminished, this has led to a decrease in their influence over the availability and mix of local services; the restriction on *Länder* infrastructure funding may have similar consequences.

The LTCI funds are also responsible, together with the care providers, for assuring the quality of long-term care services.

Eligibility and assessment for long-term care

LTCI benefits are determined by the insurance principle that people requiring similar levels of assistance because of disability or frailty should receive equal treatment. No account is taken of individual circumstances that might affect the actual levels of help needed, such as financial status or the availability of family carers.

Eligibility is determined through assessment of the duration and frequency of help regularly required with personal hygiene, eating, mobility and housekeeping. The eligibility criteria were developed to fit the estimated funds available. The amount and frequency of help needed determine the level of benefit payable, at one of three 'care dependency' levels. The eligibility criteria for each category of 'care dependency' are the same for institutional and home care and apply to those insured through both public and private LTCI schemes. Claims are assessed by medical services that are jointly financed by the sickness insurance funds and include nurses and carer members as well as doctors. Decisions about the eligibility of individual applicants are made by the LTCI fund, on the basis of recommendations from the medical service.

Nature of long-term care benefits

Benefits options are a cash payment (at a lower value); professional home care services (worth nearly twice as much); or a combination of the two. The level of the cash benefit option ranges from €215 to €675 per month, depending on the level of 'care dependency'. The cash benefit is awarded directly to the person needing care, who may then pass it on to a family or volunteer carer. The cash benefit is not treated as taxable income which, given high levels of taxation and income-related insurance contributions, is of considerable indirect value. Despite its significantly lower value, the cash option has always been much more popular than 'in-kind' services, although recently there has been an increase in the proportion of beneficiaries opting for 'mixed' awards of cash and services. In 2002 the ratio of applicants choosing cash to services was roughly 80:20.

Once eligible, the cash benefit option can be thought of as a supplement to the beneficiary's income, to be used in whatever way s/he wishes. It is widely assumed that most beneficiaries give at least some of the cash benefit to a family carer; where both share the same household, the cash benefit is likely to become part of their joint household income (Schunk and Estes, 2001). The in-kind service option 'may be thought of as a "voucher" for approved services' (Evans Cuellar and Wiener, 2000); beneficiaries can choose between any of the provider organisations that have contracts with their LTCI fund and may be able to negotiate more hours of service, at lower cost, for a given level of benefit.

For those entitled to LTCI (at any level of 'care dependency'), other benefits include:

- *Respite or holiday care* – up to four weeks a year for family carers, during which LTCI will pay up to €1,470 for substitute professional services. This benefit is also available if the usual carer is ill. The usual carer must provide at least 14 hours' care a week and have been caring for at least 12 months prior to the date of absence in order to qualify.
- *Technical aids* – special home nursing equipment and grants up to €2,557 to adapt the home.
- *Retirement pension and accident insurance contributions* – for informal carers under pension age who are employed for less than 30 hours a week and provide informal care for at least 14 hours a week – about 530,000 carers. The number of informal carers is slightly decreasing (from 575,000 in 1997–99 to 530,000 in 2001). The LTCI funds transferred per year are more than €1 billion. Some 90 per cent of carers insured by retirement pensions in this way are women (Deutscher Bundestag, 2004). Informal carers are also automatically covered by statutory accident insurance.
- *Direct support for carers* – beneficiaries choosing the cash option are visited by a nurse employed by the care insurance fund every three to six months, depending on the level of 'care dependency'. This is partly to monitor the quality of care being received and partly to provide advice and support for carers. There is no information on the acceptability or effectiveness of these visits, from the perspectives either of carers or of the LTCI funds; however, 'the prevailing view is that care among family members is a personal issue, largely outside the regulatory regime' (Evans Cuellar and Wiener, 2000, p. 21). LTCI funders are also required to offer free nursing courses for informal carers; again there is no evidence on their takeup or effectiveness. Carers are entitled to retraining opportunities if they want to return to paid employment after a period of care giving.

About 75 per cent of applicants for LTCI meet the eligibility criteria. Although the criteria are legally prescribed, there remains a persistent (though narrowing) gap between different LTCI funds in their proportions of unsuccessful applicants and their assignment of successful applicants to the different levels of care dependency (Schunk and Estes, 2001).

The LTCI eligibility criteria were initially criticised for their bias towards physical disabilities. Rather than amending the eligibility criteria or enlarging the scope of the LTCI scheme (which could have risked jeopardising its overall financial sustainability) a 2002 amendment to federal legislation (*Pflegeleistungs-Ergaenzungsgesetz* – Law on Supplementary Care Benefits) provides additional LTCI benefits to people who, in addition to the standard 'activities of daily living' (ADL) assessment, have 'continuous limitations in competences of daily life' that require constant supervision. Additional benefits consist of:

- A personal budget of €100/200 a month for the person needing care that can be spent on respite or relief care provided by a home care agency or voluntary organisation.
- Additional advice and support services costing €50 million a year for carers of cognitively impaired people, whether LTCI beneficiaries or not. The introduction of training courses for these carers represents an attempt to assure the quality of care provided to cognitively impaired older people.

The takeup of the personal budget is still not satisfactory. About 220,000 persons are estimated to be entitled to the personal budget benefit, but only 33,000 took up this benefit in 2003 (Deutscher Bundestag, 2004).

Overall distribution of resources for long-term care

The overall distribution of resources for long-term care is shown in Table 9.2. Around 70 per cent of total long-term care expenditure is publicly financed; LTCI contributions and premiums alone account for almost 80 per cent of this. The contributions from general taxation for social assistance and public investment subsidies are now of minor importance.

There is, in principle, no necessity for older people needing care to spend down their assets and therefore no impact on inheritance. However, since the levels of insurance benefits are capped by federal law, in

Table 9.2   Sources of funding for long-term care

Source of funding	As % of all spending
Public funding:	70
Public LTCI	55
Private mandatory LTCI	3
Social assistance	8
Investment financing	5
Private funding:	30
Nursing home care	23
Home care	7

Source: Adapted from Rothgang (2003).

practice private co-payments are common, particularly in meeting the costs of institutional care (although *Länder* subsidies for the construction and maintenance of nursing homes have helped keep down levels of private co-payments). LTCI benefits cover about 50 per cent of the average cost of institutional care; they exclude the costs of board and lodging and any share of capital costs. Older people unable to afford the shortfall remain (partially) dependent on social assistance. In 1999, 33 per cent of all people in institutional care received social assistance to supplement their LTCI benefit; recently this may have risen to around 40 per cent, as care costs have increased but benefit levels have not. In relation to home-based support, Thomas (2003) estimates an average monthly gap of €130 between the in-kind service benefit and the actual costs of the home care services required. Thomas also estimates that only 35 per cent of those receiving home care services manage without having to purchase additional services privately. Moreover, in 1999, about 4 per cent of care insurance beneficiaries living at home were supplementing their LTCI benefits with social assistance in order to meet all their needs for home care services; again this number is likely to have increased since (Deutscher Bundestag, 2002, 2004). Local social assistance boards use the information from the medical assessment carried out for LTCI in order to determine the level of additional help they provide. Social assistance is also available for additional communication and mobility needs.

Nevertheless, the introduction of LTCI has led to a major shift from the budgets of the *Länder* and municipal social assistance schemes and from health insurance funds, to LTCI (Deutscher Bundestag, 2004). This has also protected health insurance funds against pressures to increase their own premiums.

LTCI is a pay-as-you-go scheme. Because long-term care contributions are income-related but benefits are not, redistribution takes place from affluent to poorer contributors (and their households). Redistribution also takes place from households with two breadwinners and those with no children, to families with only one breadwinner; and from men (who are more likely to pay contributions) to women (who have greater longevity). It has also been argued that the introduction of LTCI has primarily benefited the middle classes, who previously funded the cost of long-term care out of their own private resources (Alber, 1996).

Impact on and articulation with other welfare systems

Because the line between acute and long-term care is problematic, allegations of cost shifting between sickness and long-term care insurance are common. One of the aspirations underlying LTCI, that it would stimulate the development of rehabilitation services (which are still very sparse) has largely not materialised, because the costs of these services are borne by sickness insurance while the benefits are felt by LTCI.

Supply issues

There is no funding by LTCI of care management or advocacy services; these remain sparse. One consequence of the introduction of LTCI was a significant increase in the number of home care provider organisations, from 4,300 in 1992 to 10,600 in 1999. About an additional 60,000–70,000 workers are now employed in the home care industry (2001: 190,000 workers, 86 per cent of them women); most of this growth has been in the private, for-profit sector. This has been facilitated by the LTCI legislation, which required insurance funds to favour non-profit and commercial private agencies over public agencies. However, these new provider organisations are relatively unstable and many go out of business after a short time. The supply of nursing homes grew as well, from 4,300 in 1992 to around 8,100 in 1999.

However, the LTCI funds have poor leverage over the quality of care, particularly in institutions. In most cases quality is nevertheless considered to be adequate, as shown in the first report by the sickness insurance Medical Service (*Qualität in der ambulanten und stationären Pflege 1*, 2004). But because the LTCI legislation prioritises economic over quality issues, the annual negotiations between the LTCI funds and providers focus on price rather than quality. Additional problems relate to allegations of fraud and reductions in efficiency as a result of the growth in home care service capacity; and instability among home care provider agencies in a market that remains chronically underfunded.

Recent and current debates

Pressures on LTCI are relatively small compared to those on the health and pensions insurance schemes. Nevertheless, LTCI is currently affected by wider economic pressures and by concerns about the long-term viability of those other, closely related schemes. Continuing high levels of unemployment have reduced the level of funds coming into the LTCI scheme (although the unemployment insurance funds and social assistance boards pay the LTCI contributions of their beneficiaries, these are much lower than would be paid by employed contributors).

Recent changes up to 2007 have included:

- From 2004 retirement pensioners are required to pay their LTCI contributions in full, rather than these being subsidised by their pension insurance fund.
- Social assistance boards are able to reclaim funds from the children of older claimants. In spite of the fact that a number of court decisions have decreased the financial responsibilities of families in relation to social assistance claims, the number of such claims increases as the benefits provided by LTCI are capped and additional social assistance benefits are necessary.

Two commissions of enquiry into LTCI conducted by the main German political parties reported in 2003. Both the Rürup and the Herzog enquiries concluded that no major changes should be introduced in the LTCI scheme, particularly as any changes that increased once more the role of social assistance in funding long-term care would meet considerable resistance from the *Länder* and municipalities. Nevertheless, both reports proposed reducing the higher levels of in-kind benefits currently paid to LTCI recipients in institutional care to the same level as the in-kind benefits for people living at home.

A second area of debate concerns the fragmentation and quality of services. Despite the massive injection of resources into the long-term care services market, new developments have been slow. Legislation passed in December 2001 permits long-term care insurance funds to invest resources in supporting model projects to develop the quality and coverage of services. These are time-limited projects that are subject to rigorous evaluation. One such project involves the pooling of entitlements across a number of funding systems – rehabilitation, statutory health insurance, LTCI and social assistance, for people who have complex or multiple needs and entitlements. The aim is to enhance the market leverage of the individual over the range of services s/he may need.

A second demonstration project also aims to improve users' leverage and choice vis-à-vis formal service providers, but this involves LTCI benefits alone. It is modelled on the highly popular Dutch Personal Budget scheme. Beginning in September 2004, the project aimed to recruit up to a thousand participants from across Germany, with a matched comparison group (Klie and Spermann, 2004). The LTCI higher benefits-in-kind option is combined with any private resources or social assistance to which the beneficiary is entitled and converted into a 'personal budget'. Care managers employed by the municipality then help to assemble a personalised package of services to meet the needs of the individual beneficiary. Significantly, services can be purchased from any source, not just those provider organisations with which the LTCI funds have contracts. However, resources cannot be used to give regular payments to relatives, nor can services be purchased in the grey labour market. The role of the care manager will be crucial – additional training is required for people adopting this new role.

This demonstration project is controversial and opinion is divided as to its likely impact on the current widespread preference for the LTCI cash benefit option (increased demand for the in-kind benefit option would have implications for the financial sustainability of LTCI) and on the flexibility of in-kind services.

A further area of reform is the quality of services. As the role of social assistance in funding institutional care has diminished, so older people and their relatives have become less reluctant to complain about the quality of care. LTCI has created new opportunities for care insurance funds and provider associations to discuss quality standards. The Care Quality Improvement Act 2002 assigns responsibility for quality to the owners and managers of nursing homes and home care agencies; and requires all staff in both sectors to ensure that quality standards are met. These responsibilities – and their financial implications – have to be accepted by the care insurance funds when setting contracts.

Further developments concerning LTCI in Germany

As of 2007, a reform of LTCI in Germany was in preparation, involving at least the following elements:

- raising of benefits in cash;
- indexation of financial provisions;

- raising of financial benefits for persons with cognitive impairments;
- improvement of local care provision by bundling and integrating existing ambulatory care structures (*Pflegestützpunkte*);
- introduction of individual case management (*Fallmanagement*);
- promotion of prevention and rehabilitation;
- promotion and facilitating of sheltered housing, particularly for people with cognitive restrictions;
- improvement of quality assurance;
- promoting civic involvement.

Long-term care in the UK

Recent reforms to long-term care in the UK have been piecemeal and incremental (Parker, 2000, p. 150), resulting in a complex set of arrangements for funding and accessing public resources for long-term care that have, at the same time, failed to satisfy key stakeholder groups (Robinson, 2001; Brookes *et al.*, 2002).

Raising revenue for long-term care

Funding for long-term care comes from two main sources:

- General taxation (and income-related National Insurance contributions) accounts for an estimated 65 per cent of all long-term care expenditure (Comas-Herrera *et al.*, 2004).
- Charges and fees paid by older people (and their families) for institutional, day care and domiciliary services make up 35 per cent of total long-term care revenue, of which 17 per cent comes from user co-payments and 18 per cent from the private purchase of services (Comas-Herrera *et al.*, 2004).

There is virtually no market for private long-term care insurance (Johnstone, 2005).

Organisational arrangements for long-term care

Responsibilities for health and social care are devolved to the Scottish Executive, National Assembly for Wales and Northern Ireland Assembly, with the Westminster government retaining responsibility for England. In all countries, public resources for long-term care are embedded within the mainstream budgets of the National Health Service (NHS) and local authorities. NHS services (including community-based nursing) are funded from general taxation and are generally free at the point of use. NHS resources are allocated by central government to local Primary Care Trusts (PCTs) (in England, with local variants in Scotland, Wales and Northern Ireland), which are responsible for commissioning and/or providing hospital and community health services for populations of up to 300,000 people.

Local authorities are responsible for providing personal social services, funded from a combination of local property taxes, grants from central government and user co-payments. Each of the 150 English local authorities with adult social services responsibilities covers on average 325,000 people (but they are not coterminous with PCT areas). Local authorities have some flexibility in deciding the allocation of resources between social services and other local services; however, this is circumscribed by clear statutory duties, including those of assessing population-level and individual needs for social and personal care; and by national performance benchmarks.

While many providers of health services remain part of the NHS, institutional and community-based social care services are purchased by local authorities from public, voluntary and for-profit providers. In 2001, 85 per cent of residential home places and almost all nursing-home places for adults were purchased from non-statutory providers (Means *et al.*, 2003). The volume of home care services provided by independent organisations under contract to local authorities rose from 5 per cent of contact hours in 1993 to 64 per cent by 2002 (Forder *et al.*, 2004).

Social security benefits – an attendance allowance for older people needing help with personal care and/or supervision and a carer's allowance for family carers – also contribute to long-term care resources. These are part of the UK-wide social security system, funded by taxation and administered by the Department for Work and Pensions.

Eligibility and assessment arrangements

These divisions of resources and responsibilities are reflected in the complex arrangements by which older people (and family carers) access publicly funded long-term care and the different eligibility criteria that characterise these arrangements. Up to 2002, these arrangements were largely the same across the UK, but one major difference has been introduced in Scotland.

Continuing NHS care

Since the early 1980s, there has been a dramatic reduction in long-term care provision by the NHS, although this has been the subject of repeated legal challenges by patients (Royal College of Nursing, no date). Currently, people with continuing complex, intensive, unstable or rapidly deteriorating health care needs, but who can nevertheless be discharged from acute hospital care, can receive *all* their care free of charge from the NHS – a funding regime effectively equivalent to hospital inpatient status. Eligibility is based on clinical needs alone and is decided by clinicians, in line with national guidelines.

Residential and nursing homes

Responsibility for funding nursing and residential care lies with local authority social services, along with responsibility for assessing needs and determining access to these services. Access to funding for residential and nursing homes depends on an assessment of care needs plus a test of capital/assets. Residential and nursing-home care is fully funded by local authority social services departments for older people with assets (including the value of a house) below £13,500; older people with capital of over £22,250 are not entitled to any public support for the costs of their institutional care and must spend down to this level before qualifying. (For capital between £13,500 and £22,250, a tariff income is assumed that offsets public funding on a sliding scale.) These figures are for England and slightly different amounts apply in Scotland and Wales. However, the principle is the same – a strict capital limit (including the value of housing and other property) restricts access to publicly funded institutional social care.

There are two types of institutional care – nursing homes and residential homes for those with only personal care needs. The level of care required determines the level of fee charged by the home. The NHS contributes the cost of the nursing care required in nursing homes; in England, since October 2007 this has been £101 a week. There are different levels in Wales, Scotland and Northern Ireland), depending on the assessed level of nursing needs. This nursing payment is deducted from the fee charged to an older person (or the local authority social services department that meets the fees of older people whose assets are below the capital limit).

Domiciliary and day care

Since 1993, there has been a concerted effort to increase the capacity of home care services as alternatives to institutional care. These services are now targeted towards users with more intensive or complex needs (Bauld *et al.*, 2000), although this also means that proportion-ately fewer older people receive these services. This targeting has been achieved through the use of assessment to restrict access to services (the assessment process is essentially the same as that determining access to institutional care). At the same time, there has been a shift in the *content* of domiciliary services, away from help with domestic and household tasks to a predominant focus on personal care.

There is no test of means or assets restricting access to domicil-iary or day care services. However, users in England must contribute means-tested co-payments towards the costs of these services. There remain significant variations between local authorities in the details of how these co-payments are assessed (Thompson and Mathew, 2004). Since 2002, personal care in both institutional and domiciliary settings has been free of charge in Scotland; as of 2005, Wales was debating introducing free personal care.

Most home care services are provided in kind, increasingly from inde-pendent provider organisations whose services are purchased through contracts with local authorities. However, since 2002 older people have been able to opt to receive a cash payment (direct payment) instead, to the value of the home care service they would otherwise have received. The payment is expected to be used for the private purchase of home care services or private employment of a paid carer; it cannot be given to a co-resident family carer. Although direct cash payments are popu-lar with younger disabled people, few older people have so far opted to receive them instead of services in kind.

Social security benefits

The attendance allowance can be claimed by older people needing regu-lar or substantial help with personal care or supervision to remain safe. Eligibility is assessed by a separate, standard national test of personal care or supervision needs. Insofar as the attendance allowance is taken into account in assessing co-payments for day care and home care ser-vices (Thompson and Mathew, 2004), it can be considered a transfer payment from central government to local authorities. Family carers of older people receiving the attendance allowance can also claim a carer's allowance, if they receive no other social security benefits and have no more than minimal earnings from paid work.

Assessment and eligibility arrangements – conclusions

These arrangements involve different assessment processes and crite-ria, including combinations of disability/needs-based criteria, assets tests

Table 9.3 Summary of different arrangements for accessing public funding for long-term care, UK

Type of care	Test of:			Geographical coverage of eligibility criteria	Scope of public funding		
	Need	Assets	Income		Nursing	Personal	Hotel
NHS continuing care	Yes	No	No	Country (Scotland vs rest of UK)	Yes	Yes	Yes
Nursing and residential home	Yes	Yes	No	Country (Scotland vs rest of UK)	Yes	Covered in Scotland, not rest of UK	Yes
Domiciliary care	Yes	No	No	Country (Scotland vs rest of UK) *and* local authority	Yes	Covered in Scotland, not rest of UK	No
Social security benefits (AA/CA)	Yes	No	No	UK	No	Yes	No

AA Attendance allowance CA Carer's allowance PCT Primary Care Trust.

and income-based criteria. Each combination of criteria, in turn, allows access to public funding for different aspects of long-term care. These variations are summarised in Table 9.3.

These arrangements result in several major inequities:

- Diagnostic inequities, particularly between older people with severe or unpredictable physical health problems who are more likely to qualify for NHS-funded long-term care and those with stable physical conditions or cognitive impairments;
- Geographical inequities, between older people under different local authorities who are charged different sums for the same home care services. Major geographical inequities also now exist between Scotland and the other countries of the UK.
- Intra- and intergenerational inequities, between older people entering institutional care who have to spend down their savings (including the value of a home) and those who remain living in their own homes, whose major asset of a home remains intact to be inherited by their descendants.

Recent and current debates

The Royal Commission on Long-term Care

Public concern over these inconsistencies prompted the establishment of a Royal Commission on Long-term Care. The Commission's (1999) report recommended that all personal *and* nursing care, wherever provided, should be free of charge and provided only on assessment of need. The 'hotel' (board and lodging) elements of institutional care, and 'indirect' care – help with domestic/household tasks – would continue to be means-tested.

This proposal represented a significant shift in funding responsibilities from the individual back to the state. Although it was anticipated that it would increase public spending on long-term care from 1.0 per cent to 1.4 per cent of GDP by 2051, the Commission argued this was justified by improvements in transparency and fairness. However, two members of the Commission disagreed: they argued that the costs of making both nursing and personal care free of charge could prove unsustainable; would not improve the quality of care, and would represent a regressive transfer of public resources to the better-off.

The government accepted many of the Commission's recommendations – but not that which had prompted the dissenting report. It agreed to fund nursing care in institutions, but retained the assets and means tests for personal care, at least so far as England was concerned. Instead the government opted to invest additional resources in an expansion of intermediate care and rehabilitation services to prevent emergency hospital admissions and facilitate early discharge (Secretary of State for Health, 2000).

This decision, which was implemented from October 2001, introduced a major inequity between older people in Scotland, which adopted in full the Commission's Majority Report recommendations and in April 2002 introduced free nursing and personal care in both institutional and domiciliary settings, and the rest of the UK.

Promoting well-being and independence

The transformation of traditional home help services into personal care services and the absence of almost all publicly funded help with domestic tasks (cleaning, laundry, etc.), even for the frailest older people, has been deeply regretted by older people themselves. Some influential national bodies – the Association of Directors of Social Services (ADSS), the Local Government Association (ADSS, 2003) and the local government spending 'watchdog' the Audit Commission (Audit Commission,

2004a) have all called for new investment in a wide range of local services that could sustain the independence and well-being of all older
people, not just the estimated 15 per cent minority who use long-term
care services. Underpinning this call is a concern that the wider social
needs of older people for accessible transport, recreation and leisure,
information services, housing, security in the home and neighbourhood
and so on, have become overlooked.

Increasing choice

The current government is committed to expanding choice in long-term
care. Two policy documents (Department of Health, 2005b; Department for Work and Pensions, 2005) advocated the further expansion
of direct payments and proposed an alternative means of increasing
choice – 'individual budgets'. These would bring together several different funding streams (including social services expenditure on services,
equipment and minor adaptations and other housing-related resources).
Following assessment, an older person would be able to choose how
the resources to which s/he was entitled would be spent – on conven-
tional services or on support from voluntary, community or commercial
providers. It was hoped that this would also enable older people to
access a wider range of lower-level preventive services that could help
to sustain independence and improve quality of life. Pilot individual budget projects were under development in 2005; however, many
questions arose about the information and support that older people would need to be able to exercise these new choices; about the
behaviours of providers and the availability of desired service options;
and about the lack of integration with concurrent NHS policies and provision (Glendinning, 2006). A comprehensive evaluation was due to be
published in 2008.

Evaluation and comparisons of long-term care arrangements in the UK and Germany

This final section compares five different aspects of long-term care
arrangements in the UK and Germany: eligibility of older people for
care services; support for family care giving; formal service provision;
macro-level financing; and politics and governance issues.

Comparison 1: eligibility of older people for care services

Universality vs selectivity

German LTCI is a universal scheme; virtually the whole population
regardless of age is covered, and coverage is actually wider than that of

sickness insurance. Apart from social security benefits, there is no *entitlement* to public funding for long-term care in the UK and a substantial minority of older people, in England at least, are largely excluded from public funding (apart from the costs of their nursing care) because of the assets test that determines access to public funding for institutional care. The virtually universal pooling of risk in Germany also creates a more efficient funding system.

Assessment and equity

Eligibility for German LTCI is determined by a single set of criteria, applicable across the country, that takes no account of financial status or social circumstances (particularly the availability of a family care giver) and is administered with a high degree of consistency. This feature results in high levels of horizontal equity between older people with similar levels of dependency, regardless of their particular family or economic circumstances. In the UK, criteria of eligibility for long-term care funding vary between countries; between sectors (NHS, local authorities and social security); and between care locations (institutional vs home care). There is no entitlement – legally enforceable or otherwise – to publicly funded long-term care, apart from the social security benefits of the attendance and carer's allowances. In all other sectors, eligibility is based on individualised, needs-led assessments, which can allow considerable discretion over whether wider social circumstances are taken into account, and over the levels and nature of the services provided in response to those assessments. This means that individual older people who might fall just below the threshold of a single, standardised eligibility test but who nevertheless may have additional social needs may still receive help. However, standardised eligibility criteria and assessment processes offer high levels of transparency and fairness and, as in Germany, can be linked to popular notions of citizenship. They also avoid discrimination against particular groups of older people such as those who, as is current practice in the UK, have a family carer available and are therefore less likely to be assessed as being at risk and in need of long-term care services.

Equity is also affected by the factors that are included in eligibility criteria and assessments. In both countries, older people whose care needs arise from cognitive impairments and other mental health problems tend to be disadvantaged in comparison to those with physical disabilities. The German Supplementary Law on Care Benefits 2002 did not alter the care dependency guidelines that determine eligibility for LTCI, and thus did not improve diagnostic equity – instead it provided additional benefits for those who qualify for LTCI under the standard guidelines. In England, diagnostic inequity also remains between people requiring largely nursing-related care (which is free of charge) and those (including older people with cognitive impairments) who require mainly personal help and supervision. (The introduction of free personal care in Scotland has reduced this diagnostic inequity for Scottish older people.)

Choice

Older people entitled to LTCI in Germany can choose between a cash benefit and (higher-value) services in kind. The majority of older people still choose at least part of their entitlement as a cash allowance. Those choosing the in-kind service option have further choices between those service provider organisations with whom their insurance fund has agreed the purchase of services; and can also choose the specific service interventions they wish to receive from those providers. It may also be possible to negotiate with providers to receive more service 'assignments' at lower unit cost, within the overall value of the in-kind benefit.

Fewer choices are currently available to older people in the UK. Because social services departments purchase institutional and home care services from private providers on behalf of older people, so older people receiving publicly funded long-term care can only choose between those providers with which their social services department has contracts. Care managers also tend to specify in some detail the specific tasks that service providers must undertake with an individual older person and the times at which those tasks must be carried out, leaving little scope for flexibility and responsiveness to individual preferences. Research into older people's experiences of assessment reveals little evidence of choices being offered between providers or in the content of services (Hardy *et al.*, 1999). Only a very small minority of older people receive direct payments instead of services in kind.

Moreover, choices in relation to both publicly funded and privately purchased long-term care services are increasingly constrained by problems of supply, and these constraints could also affect proposals for greater use of direct payments and individual budgets. A combination of economic factors has led to a restructuring of the nursing and residential care home sectors, with many smaller homes closing and the market becoming dominated by a small number of major provider chains (Holden, 2002). The flexibility and responsiveness of home care provider organisations is also constrained by major labour market supply problems (Henwood, 2001).

Comparison 2: support for family care giving

Long-term care in both the UK and Germany relies heavily on the contributions of family carers. In both countries these contributions are recognised, but the nature of this recognition varies markedly.

The UK is one of the few countries where informal carers have direct entitlements to financial and other forms of support in their own right, rather than indirectly through the benefit entitlement of an older person who is receiving care. Although entitlement to the carer's allowance depends on the older care recipient first receiving the attendance allowance, application for the carer's allowance is made by the carer; the allowance is paid directly to the carer; and it is treated as her/his income. This is in contrast to the German LTCI cash option, where the carer's entitlement to any help at all rests on the eligibility of the older person. Moreover, whether the cash allowance option is passed on to a family carer is also at the discretion of the older person (Glendinning and McLaughlin, 1994; Weekers and Pijl, 1998).

Moreover, informal carers in the UK have the right to assessments of their own needs by local authority social services departments and these rights are increasingly independent of the situation of the older person. Legislation has further extended carers' rights: the Carers (Equal Opportunities) Act 2004 places a duty on local authorities to tell carers about their rights and to consider carers' wishes in relation to paid work, leisure or study when carrying out assessments. The Act also gives local authority social services departments powers to involve other statutory organisations, including the NHS, in providing support for carers.

While German LTCI also recognises the contributions of informal carers, there is arguably greater emphasis on supporting their role as carers and less on recognising their own economic independence. They have no entitlement to help in their own right and their status is effectively that of a dependent of the LTCI beneficiary, whose own entitlement to benefits determines what, if any, benefits may be received by the carer. However, once the older person's entitlement is established, then secondary entitlements can be derived by the carer, including an annual entitlement to respite care up to a specified value. Carers of German LTCI beneficiaries also automatically receive generous social protection in the form of accident and pension insurance contributions.

Comparison 3: formal service provision

Service infrastructure – range and scope

The UK has a long tradition of extensive, formal, community-based health and social care services. Although the separate NHS and local authority funding streams have historically been problematic (Glendinning and Means, 2004), local inter-agency collaboration means that community-based nursing services are increasingly aligned, if not fully integrated, with local authority social care services. Since 1993, local authority social services departments have had lead responsibility for funding and assessing needs for all non-health long-term care, whether provided in institutional or community settings. This has enabled them to restrict institutional care only to the most severely disabled older people; and to shift the focus of the home care services they commission from domestic and household tasks to personal care. Together these strategies have meant that increasing numbers of very frail older people are able to remain living in their own homes, rather than entering institutional care.

In Germany the widespread popularity of a LTCI cash benefit partly reflects the fact that, because very little funding had previously been available, domiciliary and day care services were scarce and fragmented. The cash allowance was therefore intended to support and sustain the family-based care which, for many older people, was the only alternative to institutional care. The introduction of new resources in the form of the LTCI in-kind service option has led to some expansion in the capacity of home care services, including help with domestic and housekeeping tasks. However, it is not clear that these measures have been sufficient to reduce the dominant role of institutional care within the range of formal service options; between 1996 and 2003 the numbers of care insurance beneficiaries in institutional care actually rose from 384,562 to 613,274, whilst the number of beneficiaries of home care was more or less stable (at about 1.36 million) (Deutscher Bundestag, 2004). Both the Rürup and the Herzog committees of enquiry recommended further measures to discourage the use of institutional care by harmonising benefits between the two sectors and removing the higher levels of benefits payable for institutional care.

Coordination of long-term care services

In England, local authorities play a major role in planning and commissioning institutional and home care services; increasingly these activities also involve local health services too. Joint funding and provision of intermediate care services (see below) by health and social services organisations are also widespread.

At an individual level, coordination is facilitated by care managers who act as micro-purchasers, putting together individualised 'packages' of services. The care management role is usually carried out by trained social workers but increasingly occupational therapists and district nurses also play this role. The introduction of 'individual budgets' is likely to require a transformation of care managers' role, from micro-purchasers to brokers and advocates on behalf of older service users.

In Germany, planning mechanisms to ensure that services are available to meet local needs remain embryonic. Given the dramatic reduction in social assistance funding for long-term care, the role of the *Länder* and municipalities in long-term care planning and provision has actually diminished. It remains to be seen whether experiments with care management will achieve better outcomes for LTCI users and what the impact will be on the wider formal service system.

Boundaries between health and long-term care

The UK government's justification for not adopting in full the recommendations of the Royal Commission on Long-term Care was that resources would be better invested in short-term, intensive health and social care services to prevent emergency hospital admission and facilitate early discharge (Secretary of State for Health, 2000). Considerable investment has subsequently taken place in nursing, personal care, physiotherapy, speech and occupational therapy services and a variety of short-term intensive support, convalescence and rehabilitation units have been created. Normally organised within the NHS, these services are free of charge for up to six weeks; after that the usual long-term care assessment and funding arrangements come into play. However, disagreements remain over the longer-term effectiveness of intermediate care in improving outcomes for older people, as distinct from organisational outcomes such as reducing length of hospital stay (Means *et al.*, 2003).

The separation of LTCI from sickness insurance in Germany has, in contrast, offered few incentives to develop rehabilitation or preventive services or to improve the integration of health and social care provision. Indeed, the costs of these services are met by sickness insurance, while any benefits (in terms of reduced numbers of eligible claimants) are felt by LTCI. The federal government clarified the respective responsibilities of the two insurance schemes, in order to improve preventive services. In its 2005 report, the national commission of health system experts came out in favour of an integration of sickness and long-term care insurance in order to avoid inefficiencies (Sachverständigenrat zur Begutachtung der Entwicklung im Gesundheitswesen, 2005). The sickness insurance scheme still funds home care for four weeks, where this is part of a programme of prescribed medical treatment, but only if the older person is not eligible for LTCI and there is no informal carer available. It now also funds home-based nursing, personal care and domestic help, in addition to medical treatment, for severely disabled people for four weeks, where this reduces length of stay in hospital or while waiting for hospital admission. From 2005, sickness insurance took over responsibility for the medical care needed by nursing-home residents.

Price competition

The German LTCI legislation has offered opportunities to break the strong monopoly exercised by the 'cartel' of the major associations of service providers. Annual negotiations now take place in which service prices are agreed with the purchasers, the LTCI funds. The legal requirement to give preference to charitable and for-profit providers in these negotiations has also increased competition.

Similar negotiations take place at local levels between English social services departments and local service providers. However, severe funding constraints on local authorities have restricted the levels of fees they are able to offer and this has placed serious financial pressures on the residential and nursing-home sector in particular. Consequently, some homes have closed, reducing the choices available for older people in many localities.

Improving quality

Long-term care services in the UK are increasingly subject to a range of central-government-imposed quality benchmarks and inspection regimes. The National Service Framework (NSF) for Older People (Department of Health, 2001b) sets out eight standards for health and social care services. These relate to age discrimination; person-centred services; intermediate care; the quality of hospital care for older people; the prevention and prompt treatment of strokes, falls and mental health problems in older people; and the promotion of health and active life in old age. The NSF has the potential to make a significant difference to older people's experiences of health and social care services. Whether it can do so without additional resources is unclear – any such resources have, so far, only been directed at intermediate care.

Following the recommendations of the Royal Commission on Long-term Care, a National Care Standards Commission (NCSC) was established in 2002, responsible for publishing standards and inspecting, *inter alia*, care homes, home care providers, adult day-care centres, nursing agencies and private and voluntary hospitals and clinics. In April 2004, the NCSC was replaced by the Commission for Social Care Inspection (CSCI), which is also responsible for inspecting local authorities' in-house home care services and residential homes.

Lacking this central government involvement, the ability of German LTCI to improve the quality of care is relatively weak. Nevertheless, the introduction of LTCI has for the first time created opportunities for debate at federal level about the quality of care. From 1995, LTCI funds were required by federal government to develop guidelines for assuring the quality of services provided by the organisations with which they set contracts; further legislation in 2002 assigned responsibility for quality to the owners and managers of nursing homes and home care agencies. Furthermore, the decreased reliance on social assistance funding has made older people and their families less reluctant to complain about poor quality care, particularly in nursing homes.

However, new rights in Germany still focus predominantly on older people as individual 'consumers', with less emphasis on establishing and maintaining universal standards of care. This suggests that both cultural and organisational changes are still required in order to make quality of care a mainstream policy issue. Although Germany is not currently experiencing the shortages of care staff that exist in the UK, there are nevertheless problems in obtaining care staff with appropriate qualifications (especially in dementia care) and training and qualifications frameworks are fragmented. Moreover, in both countries, resource constraints mean that little money is available to invest in staff training or in additional social activities to improve the quality of life of nursing-home residents.

The extensive and explicit reliance on informal care within German long-term care funding arrangements also restricts quality measures. Given that supporting older people is widely regarded as a private, family responsibility, policies to safeguard the quality of family care are embryonic, despite the very substantial state investment in supporting this form of care. Insurance beneficiaries choosing the cash option are visited by a nurse employed by the care insurance fund every three to six months, partly to monitor the quality of care being received. However, no information is available on the acceptability or effectiveness of these visits, from the perspectives of either carers or the LTCI funds. LTCI funders are also required to offer free nursing courses for informal carers; again there is no evidence on their takeup or effectiveness. The 2002 Law on Supplementary Care Benefits introduced additional advice and support services costing €20 million a year for carers of older cognitively impaired people, whether LTCI beneficiaries or not. The introduction of training courses for these carers represents an attempt to assure the quality of care provided to cognitively impaired older people.

Comparison 4: macro-level financing

Figures 9.1 and 9.2 show that the two countries devote similar overall proportions of their respective GDPs to long-term care; exhibit similar ratios of public to private financing of health and long-term care; and allocate similar proportions of total public expenditure on long-term care between home-based and institutional services.

However, despite these similarities there are major differences in funding arrangements between the two countries. In the UK, resources for

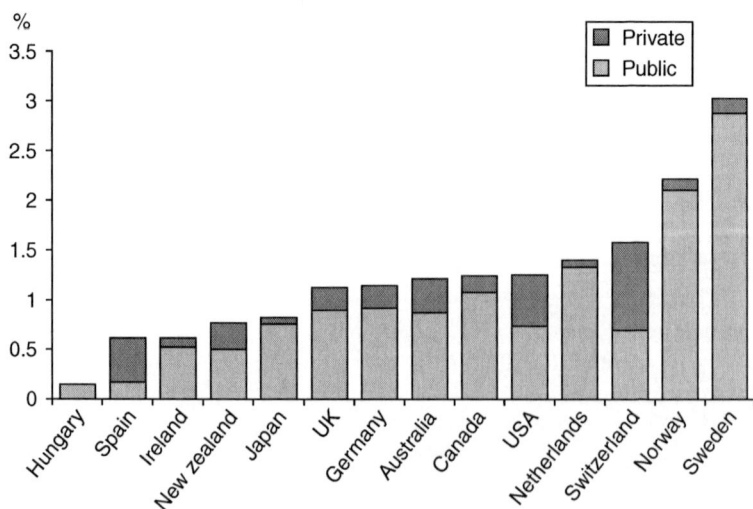

Source: OECD (2005a), p. 26.

Figure 9.1   Public and private expenditure on long-term care, selected countries, 2000 (% of GDP)

%
3
2.5
2
1.5
1
0.5
0

■ Home care (including services in
support of informal care)
□ Care in institutions (nursing homes
and the like)

Hungary Spain New zealand Ireland Luxembourg Switzerland USA Japan Austria Australia UK Germany Canada Netherlands Sweden

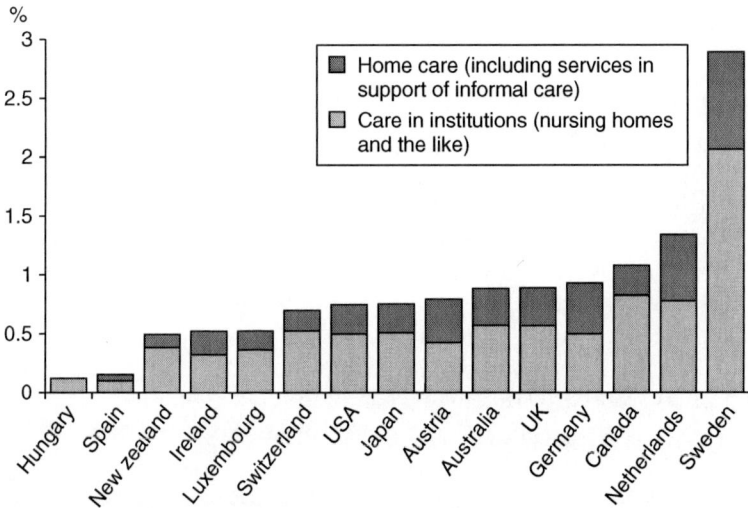

Source: OECD (2005a), p. 28.

Figure 9.2 Public expenditure on long-term care, selected countries, 2000 (% of GDP)

long-term care are embedded within mainstream NHS, local authority and social security budgets; piecemeal, incremental and sometimes
covert changes have taken place within each sector and often in isolation from other sectors. This has created a highly complex set of
arrangements for accessing services that seriously compromise equity.
Moreover, the fragmentation of long-term care resources in the UK may
have implications for the sustainability of funding arrangements, as
'ring-fenced' budgets may be easier to control (Glendinning *et al.*, 2004).

In contrast, the creation of LTCI in Germany, albeit consistent with
the insurance-based structure of German welfare, has been widely popular and largely free from concerns about its longer-term financial
sustainability. Universal coverage has been achieved, accompanied by
high levels of horizontal equity. Federal legislation sets levels of contributions, the guidelines that determine access to benefits and the levels
of those benefits. None of these parameters can be changed without
amending legislation. However, the failure to increase levels of benefits
since the introduction of LTCI is leading to a growing shortfall between
benefits and the actual costs of care, particularly in institutions, so at
least partial dependency on means-tested social assistance is growing
once more.

German LTCI also recognises, rewards and provides social protection for family carers. This constitutes a further important cost containment mechanism.

Comparison 5: politics and governance issues

Underpinning these contrasting funding regimes are major differences in the responsibilities of the UK and German governments for long-term care policy. The German reforms have substantially shifted responsibility for long-term care from the *Länder* and municipalities to the federal government, which retains legislative control over all aspects of the LTCI scheme. Moreover, LTCI has created new opportunities for discussion and debate at federal government level about wider long-term care policy issues such as the quality of services and rehabilitation. In the UK, such forums remain fragmented between territorial governments, the NHS and local authorities.

Debates over the levels of public resources allocated to long-term care in the UK and the balance between public and private responsibilities have not been resolved. In 2001, an extensive review of NHS spending and performance was conducted (Wanless, 2001). This led to unprecedented real-terms increases in NHS funding, in order to narrow the gap between UK health spending and the average for other European Union countries as a proportion of GDP. However, the government has resisted repeated calls from older people's interest groups to carry out a similar review of social care needs. Instead this was conducted by the King's Fund, an independent policy unit; meanwhile a charitable body, the Joseph Rowntree Foundation, has attempted to generate policy debate about alternative long-term care funding options. There remains considerable evidence of substantial underfunding of long-term care, with older people denied or forced to wait for help they need (Help the Aged, 2002).

In contrast, public support for a stronger role for state funding in long-term care remains high (Deeming and Keen, 2003). In particular, there is a view that an implicit contract between the state and current generations of older people, that their long-term care needs would be met, has been unilaterally broken (Robinson, 2001). The enforced sale of an older person's home under the assets test that determines access to public funding for institutional care remains a source of considerable resentment (Brookes *et al.*, 2002). In contrast, the decision of the Scottish Executive to fund personal care from general taxation provides a lever for campaigners elsewhere in the UK to press for more generous funding.

Conclusions

Comparisons of the UK and German approaches to long-term care reveal little evidence of policy convergence. Rather, developments in each country are characterised by strong elements of path dependency – each has adopted policies that are consistent with previous structures, organisations, funding arrangements and cultures. This offers major opportunities for comparisons and debates. The most significant areas of contrast are:

- national assessments based on a single set of eligibility criteria vs individualised, conditional assessments that allow scope for professional discretion and lead to regional variations;
- universal entitlements linked to concepts of citizenship vs discretionary professional judgements;
- a single, ring-fenced budget and a rational approach to pricing and costs, closely controlled at national level vs funding embedded in mainstream health and social services, quasi-market relations between providers and purchasers and poor coordination between the different funding streams;
- explicit recognition of the work of family carers vs provision biased against family carers;
- dependency of family carers on the person receiving care vs individual rights and entitlements for family carers;
- fragmentation of services at regional and local levels vs locality planning and service-commissioning mechanisms;
- fragmentation of services at an individual level vs extensive care management services;
- market-type relationships between purchasers and providers which contain weak levers for improving quality vs market-type relationships between purchasers and providers in which central government sets and enforces quality standards.

Despite these profound, systemic differences, both countries are now experimenting with projects that involve pooled funding streams to maximise the resources available to individual older people with complex needs for the creation of integrated, individualised care packages. In Germany, projects have experimented with the introduction of care managers to help put together flexible, individualised packages of care. Proposals in the UK involve a shift in responsibilities from care managers to individual older people who are promised greater choice over the nature and content of the support they obtain through an 'individual budget'. In both countries, these changes raise new questions of equity, quality and supply. It remains to be seen whether these developments will eventually become part of mainstream long-term care arrangements; if they do, this could represent a profound convergence between two currently very different approaches to long-term care.

10
Societal and Political Participation

Tony Maltby and Christiane Rohleder

Introduction

This chapter presents evidence of societal and political participation in Germany and the UK and the different approaches that the two countries choose to foster participation of older people. First, it should be noted that the political structures of the two countries are very different, with Germany presenting a federal structure where the *Länder* (the regional tier of government) and the communities have different responsibilities to implement local and national initiatives. The UK is still a constitutional monarchy with no federal structure, although it has embarked on the devolution of some legislative powers to Scotland and Wales (where there are considerable differences in policy toward older people than in the UK as a whole). The electoral system of Germany is also very different to that of the UK, being based around a system of proportional representation (PR) whereas that of the UK still relies upon a simple-majority system. These differences have influenced to varying degrees the type of social and political participation among older people and the nature of the policy-making process. Second, it has become very clear that the question what societal and political participation of older people means in reality is answered very differently in the two countries. This difference has been one of the main challenges for our comparison.

The chapter will use a similar framework to the other chapters and will focus on policy initiatives since 1990. We conclude by suggesting that with the ageing of the population in Europe there is greater potential for a new political relationship with older people, especially with policy development in the European Union at Commission level being influenced by an 'active ageing' framework. This is not only a result of the ageing of the population in Europe (making older people a numerically dominant political force) but also a result of the effects of past policies yielding more proactive, better educated, hence more demanding senior citizens.

Background

In the UK, the election of a Labour government in 1997 brought greater emphasis on the need for closer cooperation in the delivery of public services between central and local government departments, non-governmental organisations (NGOs) and importantly (senior) citizens within its 'Modernising Government' agenda (see for example Department of Employment, Training and Rehabilitation (DETR), 1998) and its later document *Winning the Generation Game* (Cabinet Office, 2000). In the former it sought to move away from the top-down model of policy implementation to a more empowering and participative framework. In the latter, it sought to attempt 'to turn the so-called burden of old age into an opportunity' (Biggs, 2004, p. 102) through, among other initiatives, the encouragement of older people to work as volunteers. Together these developments saw a growing need to place the 'service user' as citizen at the centre of policy in a way that is reflected across the whole spectrum of current social and public policy, and in particular that towards older people (defined as aged 50 years and over). In terms of implementation we can discern from as early as 1998 the development of a 'compact' between the statutory and voluntary sectors and greater 'partnership working' encouraged. Although the aim of this latter strategy has been to encourage a more equal role in policy making between sectors, this is the subject of academic debate. The process of participation is as important as the (policy) outcomes (see Barnes, 2005; Lewis, 2005).

It is suggested here that one of the principal drivers for the development of this policy grew out of a realisation that the policy achievements of successive governments, involving the development of a 'welfare state', had resulted in better educated, healthier and more active senior citizens, together with a realisation of the social and political capital that an ageing population could attain (see Walker, 1998). Cynically, we might also add that politicians saw the need to incorporate these senior citizens into the political system as a result of their greater proportion in the population. In short, we see from the mid 1990s the development of what Walker (1998, pp. 17 *et seq.*) terms 'a new politics of old age'. This, he argues, is expressed in an upsurge in 'direct political involvement' through the formation of campaigning and single-issue pressure groups such as the National Pensioners Convention (NPC) in the UK (Walker, 1998). At central government level it can be recognised in Whitehall by the early establishment of the Inter-Ministerial Group on Older People and a newly discovered enthusiasm for political lobbying on the part of the two main campaigning charities, Age Concern England and Help the Aged. At the local level in the UK, the main focus nationally for societal and political participation has been through the 'Better Government for Older People' initiative, and mirroring this and more recently, the steering group of the Older People's Research Programme (2000–04) which was comprised of older people drawn from Older People's Forums. We discuss these two developments later.

Turning our focus towards Germany, with regard to the better circumstances of each new generation of older people, the necessity to foster social and political participation was discussed from the beginning of the 1990s. In different political statements, starting with the Federal Plan for Older People from 1992, to the present action programmes of different *Länder*,[1] to the first statement of the national expert group for the regional implementation of the Madrid International Plan of Action on Ageing (Expertengruppe Nationaler Aktionsplan, 2004), the necessity to foster the societal and political participation of older people in Germany has been stressed. Key considerations in this context are the legal coverage of the senior citizen advisory boards, user involvement of older people in regard to the development and quality assurance of care services and institutions, education and new media in old age, and the promotion of volunteering and civic commitment of older people (Bundesarbeitsgemeinschaft der Senioren-Organisationen, 2004, pp. 10ff).

Whereas the discussion about user involvement can be seen as underdeveloped in Germany until now (Naegele and Walker, 2002a, p. 13), the political promotion of activity and productivity in old age has become one main driver in a whole range of policies. The chief assumptions behind this discussion are that because of the enormously increased lifetime after retirement and the wish of older people to participate actively and seek self-determination in their social activities and lives, there is a strong social need for meaningful and culturally accepted new social roles for older people (Enquete-Kommission 'Zukunft des bürgerschaftlichen Engagements', 2002, p. 213). Recently, many political activities at federal, *Länder* and local levels deal with this aim.

Additionally, as Walker (1998, p. 20) notes, such shifts are not confined to the UK and Germany alone. There is a similar shift towards participation all across Europe. For example, local authorities from

Sweden to Italy and France to Austria have established advisory boards of senior citizens. The development of this shift in terms of provision of services to older people and their involvement and participation in delivery of services was supported by the work of a number of individuals (see, for example, Osborn, 1990, 1991; Allen *et al.*, 1992; Beresford, 1992, 1997; Thornton and Tozer, 1995; Barnes, 1999).

This increased participation not only had the central aim of achieving more responsive and sensitive services, but also involved a politically more legitimate social policy arising out of widely reported demographic shifts. In both countries this shift towards a population profile which is ageing is referred to less in policy documents and general discussion in terms of 'burdens' or 'time-bombs', but increasingly as a positive achievement of a 'welfare state'. Additionally, policy documents increasingly refer to the collective potential of older people, as a resource, to solve the challenges of these future demographic changes.

Recent trends and policy debates

One of the main difficulties of this comparative analysis of the socio-political approaches of both countries has been a terminological one. This has focused around the question of what participation means and how it is interpreted and understood in the two societies. In English, often the words 'involvement' and 'participation' are used interchangeably; other words such as 'engagement' and 'consultation' have also been employed in this context (Thornton, 2000). Yet they mean very different things to different people. As Thornton establishes, little research has been conducted that has asked service users directly what these 'words mean to them and which they prefer' (*ibid.*, p. 1). Within this chapter and for the UK we use 'participation' to imply incorporation into the policy formulation and at times implementation processes. Thus, there has been seen a requirement to accommodate service users, and older people more generally, into the decision-making processes, as we noted earlier. Indeed, in the UK context, the emphasis on greater levels of participation being placed by central government has led to participation coming often to be viewed by many as an 'add-on'. There is a general view that there is a surfeit of such 'consultations' being conducted yet there has been little by the way of coordination of the valuable results and important policy lessons they report (Thornton, *ibid.*). This has led in some local instances to another perception on the part of older people (and indeed other service users) that many such participatory actions and attempts at involvement are merely paying 'lip service' to such hard-fought views or 'voices' (see Cook *et al.*, 2004).

They suggest that policy makers may hear but often do not listen, understand or wish to act.

Indeed, the 'Better Government for Older People' (see below) evaluation (Whitton, 1999) refers to the tendency to look for 'quick wins' in an effort to influence policy and not to focus upon what the real concerns of older people are. Carter and Beresford (2000) take up this point when they outline the strengths and weaknesses of different forms of involvement: the consumerist and democratic approaches. The former implies that service providers view older people as consumers or customers and participation is fundamentally about information gathering in order to gain some understanding of the service so that providers can implement strategies and services. Carter and Beresford (*ibid.*) use the analogy of market research in this context. The consumerist approach is linked to a top-down model of decision making, with control over any decisions residing with the policy maker or service provider. The democratic model argues for a shift in this power dynamic to facilitate people having a say in decisions and what happens to them: the transfer of power. It is truly a 'bottom-up' approach to policy making. They cite one example from their research to demonstrate the difference between the two models. A service provider asks the question, 'What do you see as the main health problem for older people in this area?'; the reply from the older people involved in the health campaign was one word – 'transport' (Carter and Beresford, *ibid.*, p. 16).

In Germany the discussion about participation in old age is focused on slightly different aspects. User involvement is a relatively new topic yet two major political achievements to strengthen the position of older people as users can be mentioned here. The first is the amendment, in 2002, of collaboration in stationary care, which strengthens the influence of advisory boards in old people's homes. The second measure is the Care Quality Assurance Law of 2001 that aims at more transparency and protection for users of stationary and non-stationary care. Both initiatives concentrate more on user protection than on user consultation. On the whole, the discussion about better user involvement in Germany is at an early phase and just beginning.

While the discussion about political participation in Germany is concentrated on the role of senior citizen advisory boards in communities, the main emphasis in the discussion about societal participation in old age is not so much on the question of participation in decision making as on integration into social relationships and social activities. With regard to the growing number of older people, 'successful ageing' becomes more and more connected to the concept of 'productive ageing' (Knopf *et al.*, 1989, 1999; Baltes and Montada, 1996; Künemund,

1999, 2000), where old people contribute their knowledge, their abilities and their time in socially useful ways, with the effects of higher individual satisfaction, training of abilities and social benefits at the same time. In this perspective, the social development of the potential of older people is seen as a political duty, because of its function as an 'age prophylaxis' (Baltes, 1987). Besides this, the financial crisis in public budgets, the underlying limits of the German welfare state in regard to the social support of older people, and the problems in the labour market (especially the rising number of long-term unemployed aged over 50) are reasons why in Germany the promotion of volunteering in general (Rauschenbach, 1999, p. 7) and especially in old age (Rohleder and Bröscher, 2001, pp. 23ff) is seen as one solution to the present and future challenges of demographic and social change. This refers to a much wider conceptualisation of societal and political participation in the sense of civic engagement or *bürgerschaftliches Engagement*.

Returning to the UK, in England a special analysis (Office for National Statistics, 2000) was undertaken for the 2000 General Household Survey (GHS) which used civic engagement as one aspect of social capital, following the work of Putnam (2000) and others. Civic engagement here was understood as the degree to which people participate in community life and the extent to which they feel empowered to change their society. This ranged from reading local newspapers, to writing letters to the papers, to the way in which individuals can work to solve community problems. So the definition was again broad, less specific and, we would suggest, too inclusive. Again, a less recent indication (but the most recent at the time of writing) of the levels of societal participation, contained in the 1997 National Survey of Volunteering in the UK (Davis Smith, 1997), confirmed previous research. It found that those in work and from the higher socio-economic groups were more likely to be volunteers, with the level of volunteering peaking in middle age (see Table 10.1). There had been a slight decline in formal and informal volunteering and there were more retired people involved in volunteering activity than in the previous survey six years earlier. That said, there was a higher preponderance of volunteering among cohorts aged 25 to 50. The definition of volunteering used was also very broad, and given as:

> Any activity which involves spending time unpaid, doing something which aims to benefit someone (individuals or groups) other than or in addition to close relatives or to the benefit the environment. (Davis Smith, 1997)

Table 10.1 Level of volunteering by age, the UK, 1981–97 (% of respondents)

Year	18–24	25–34	35–44	45–54	55–64	65–74	75+
1997	43	52	52	57	40	45	35
Base	111	296	284	228	193	200	174
1991	55	60	63	60	46	34	25
Base	171	301	265	193	203	217	133
1981	42	52	60	48	33	29	–
Base	258	339	289	310	283	327	–

Source: Davis Smith (1997).

There is also another issue at stake here, with many voluntary organisations applying upper age limits for their volunteers, typically 65 or 70, as there is considerable difficulty in obtaining insurance cover for their activities. So there is some ageist practice, restricting older people from volunteering, and this is one of the reasons why surveys such as the two mentioned do not demonstrate high levels of voluntary activity among those over 50.

Returning to the analysis conducted for the GHS 2000 and the much wider measure of civic engagement, two summary indicators of such engagement were used. The first indicated whether older persons felt informed, could influence decisions or were involved, for example, in collections for charities locally. When looking at these data by age and sex, civic engagement was lowest among the 16–29-year-olds and increased with age up to the 40–49-year-old group. There was little variation with age among the older age groups (Figure 10.1).

Taking action to solve a local problem showed a decline among the oldest age group, being highest for the 50–59-year-olds but at only 34 per cent. In other words, two thirds (66 per cent) were *not* involved; and we should perhaps attempt to investigate why this was.

In comparison to England, other European countries and the USA, the discussion about civic engagement in Germany is relatively new. Although we find a broad tradition of volunteering in Germany, especially in local self-government and in clubs and associations, the self-organisation of citizens to manage public affairs is not a constitutional part of the concept of German citizenship (Enquete-Kommission 'Zukunft des bürgerschaftlichen Engagements', 2002, pp. 90ff). In the German tradition of nationality, the state, not the citizen, is responsible for social welfare. Additionally, because of the close cooperation between the state and welfare associations, which were often constituted

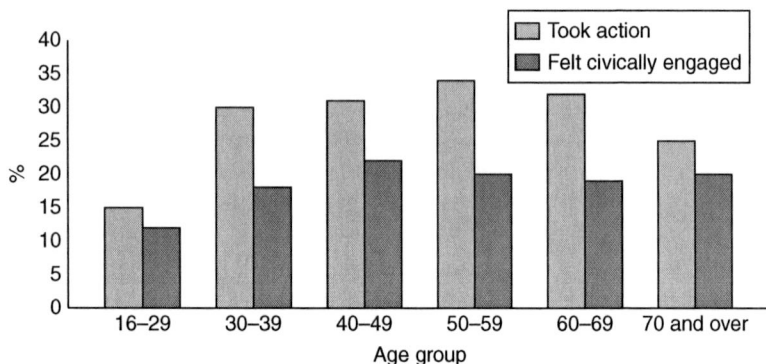

Figure 10.1   Whether took action to solve a local problem and whether felt civically engaged, by age, Great Britain, 2000 (% of respondents)

by way of volunteering, the German understanding of civic commitment is limited to social volunteering.

An important impetus to a wider conceptualisation of civic commitment was made by the Commission of Enquiry 'Towards a Civil Society with a Future' that was initiated by the Bundestag in 1999 and published its extensive report in the year 2002. In this context the first national 'survey of volunteers' was carried out in 1999 and replicated in 2004. Volunteering is here defined as work or a task that is voluntarily carried out by a person or persons as member(s) of an organisation, an association, a project or a self-organised group (Rosenbladt, 1999, p. 43). This conceptualisation is similar to that used in the National Survey of Volunteering in the UK. And the results of the first survey concerning the extent and the social structure of civic commitment of older people are similar too (Brendgens and Braun, 2000). First, as in England, civic commitment in old age is socially selective, based on gender, education and income (Rohleder and Bröscher, 2001).[2] Second, the data revealed that most of the older volunteers work in age-unrelated associations and fields despite the great political attention to self-organised initiatives for older people.

Although the results in Germany and England concerning the structure of civic commitment in old age are similar, the consequences are not – in Germany the need to foster the assumed and unused potential of seniors became an important political target. The political efforts seem to have been successful – the replication of the survey of volunteers in 2004 shows that in the intervening five-year period the percentage of active seniors aged 56–65 rose from 34 per cent to 40 per cent, and in the age group 66–75 from 26 per cent to 31 per cent. At present the so-called 'young old' are as much committed to volunteering as the age group 36–55, and in comparison with all other age groups the young old aged 55–75 showed the most distinctive increase in commitment (TNS-Infratest, 2004). It seems as if the 'unused potential' was activated.

Policy initiatives

This section outlines the more recent initiatives in each country separately, then offers an analysis of differences and similarities. In both discussions we focus upon the major policy initiatives but it must be recognised that these have encouraged and will encourage other small-scale projects to develop at the local periphery.

The German perspective

As background information, it is important to know that there is only a weak legal claim for the promotion of social participation in old age in Germany. Budget cuts in this field are common because of the public financial crisis. Secondly, the promotion of social participation is first and foremost the task of the *Länder* and the local authorities. Because of the great importance which is ascribed to civic commitment in Germany as a strategy to promote social participation, the focus of the following analysis is upon the main political strategies in this area. As there is no consensus in Germany about the best strategies to foster civic commitment (Olk, 2003, p. 307), we find various initiatives of the *Länder* and the local authorities. The role of the federal government is to initiate pilot schemes for the municipal level, to support nationwide networks and working groups, to finance research and public relations and to improve the legal conditions for volunteering (Enquete-Kommission 'Zukunft des bürgerschaftlichen Engagements', 2002, pp. 606 ff). Therefore we find the federal government providing an important impetus through various pilot schemes in this area too, which will now be discussed.

Federal and Länder pilot schemes to foster social participation in old age

In comparison to other European countries, in Germany the local infrastructure to promote volunteering is still in its early stages. One important policy scheme (Modellprogramm Seniorenbüro), intended to establish a local infrastructure for activity and civic commitment in old age, was supported from 1992 to 1997 by the Federal Ministry for the Family, Seniors, Women and Youth. This programme promoted the building of a network of small-scale local information and advice centres for senior citizens. The development of the profile of these 'senior citizens' offices' reveals a shift in the political discussion concerning the question of 'successful ageing'. In the beginning of the programme the offices were meant to encourage seniors to take up activities, for example education, sports, travel, as well as volunteering (Klages, 1999, pp. 13ff). The political intention to support 'active ageing' had been stressed in the first interim report of the Commission of Enquiry into Demographic Change (Deutscher Bundestag, 1994, pp. 340ff). Today the main priorities of the senior citizens' offices reveal a strong orientation to the promotion of civic commitment (Braun and Emons, 2000, p. 20) and through that a shift to the topic of 'productive ageing'.

Today there are 160 senior citizens' offices throughout Germany. Their interests are represented by the Federal Working Group of Senior Citizens' Offices (BaS), established in 1995. While on the one hand the senior citizens' offices are important local information centres for older people, it has to be stated, on the other hand, that their professional profile is not uniform: they differ extremely in their targets, in the structure of the agency and in their financial and personnel resources (Enquete-Kommission 'Zukunft des bürgerschaftlichen Engagements', 2002, p. 308). Because of the financial shortfalls in the communities their continued existence is not assured.

The main programme of the Federal Ministry for the Family, Seniors, Women and Youth, dating from 2002 to 2006, is called 'Experience-based Knowledge for Initiatives' (EFI – Erfahrungswissen für Initiativen) (Braun *et al.*, 2004). The programme runs in ten federal states and in 35 communities. The objective of the programme is to offer new social roles for senior citizens by training them to become a so called *SeniorTrainerin*, a 'multiplier' for civic commitment. The training will enable senior citizens to support the idea of civic commitment in the community, to establish new groups and to consolidate existing initiatives. The programme is tied to the so-called 'agencies for civic commitment' – local offices for senior citizens, centres for volunteering and advice centres for self-help. They select the participants and are responsible for their further support. The participants in the programme are in the majority, with an average age of 61, yet their educational and professional background is above average. Most of them were already volunteering when they joined the programme. This refers to a problematic point of the present strategies to foster social participation – the question whether they mainly reach privileged senior citizens.

Current ideas for a new federal programme are connected to the discussion of the abolition of universal compulsory military service in Germany and with it the abolition of community service as an alternative to military service. One result of this abolition will be a shortfall of staff in many charitable institutions. Because of this development the report of the Commission on Impetus for Civil Society (Kommission Impulse für die Zivilgesellschaft) (2004) recommends the extension of the already existing Voluntary Year of Social or Environmental Service that today is only directed at young adults. Because of the decreasing number of young people, the concept of the Voluntary Year will be adapted to the interests of senior citizens and new fields of commitment. This is seen as a further step to the integration of older people into new, responsible social roles. But it can also be interpreted as a further step to a social re-obligation of senior citizens with regard to the challenges of the demographic shift, the problems of the labour market and the welfare state.

Over the past years there have, additionally, been different special programmes in the federal states to foster volunteering. These show conspicuous differences between East and West Germany. In West Germany the programmes deal with the topics of the underlying limits of the welfare state, the promotion of social capital and the need to change the relation between citizens, local authority and politics (Braun, 2003, p. 147). The objective of the programmes is to foster the self-help potential of older people, for example in the form of one of the first pilot schemes in this area, the project established in Berlin in 1986 on 'How to Use the Experience-based Knowledge of Older People' (Müller, 1996). Other projects are the so-called 'Senior Citizens' Cooperatives' which were established in Baden-Württemberg in 1990 (Otto, 1995),[3] or programmes such as 'ZWAR – Between Work and Retirement' (Klehm, 2002) and 'Network Düsseldorf' (Nell, 2000) in North Rhine-Westphalia that concentrate on the building of strong networks between older people as a protection against social isolation and loneliness in old age.

In East Germany, programmes are more connected to the problems of the labour market (Braun, 2003, p. 147) because of the high numbers of older people who are long-term unemployed. Examples are two programmes in Sachsen: 'Action 55 – Sachsen Needs You (Baur *et al.*, 1997) and 'TAURIS' (Zur Initiative TAURIS, 2002), the latter supported by the European Union. The special feature of these programmes is the focus on younger senior citizens, already retired or long-term unemployed, who are placed by negotiation with volunteer organisations and who get a tax-free expense allowance as an acknowledgement of the low income of old pensioners in East Germany.

Generally we can identify in Germany several priority shifts in the political strategies to foster civic commitment in old age:

- In the beginning, the negotiated placement of older people in voluntary associations and organisations was the main target. Today the training of 'multipliers' to support or establish initiatives and promote the idea of civic commitment has gained in importance.
- The importance of training and qualification as an incentive for older people to volunteer has increased.
- Cross-generational commitment and a greater awareness of age diversity have gained in importance.
- The present strategies emphasise the performance and potentials of older people and try to create new, socially appreciated roles for old age.

All strategies to foster social participation have to deal with the change in the motivations and expectations of older volunteers. The promotion of civic commitment is not free of charge. Senior citizens are increasingly interested in voluntary tasks that allow participation in decision making and contact with other volunteers, offer flexible time management, and bring social appreciation. Last but not least, the demands on sufficient and unbureaucratic financial support of voluntary work will increase (Brendgens and Braun, 2000, p. 263; Rohleder and Bröscher, 2001, p. 35). However, the local administration structures are not adjusted to these demands.

The overall representation of political interests in the field of senior organisations rests with the Federal Working Group of Senior Organisations (BAGSO). It is a loose union of at present 86 organisations working for and with seniors. The senior organisations of the political parties and trade unions are members as well as welfare organisations, self-help organisations in the field of health care, church organisations and a variety of small local initiatives. Whereas the Commission of Enquiry 'The Future of Civic Activities' stressed the possibilities of this heterogeneous profile, another view is that although BAGSO functions as the national contact concerning political questions of old age, it has no democratic authorisation and its orientation is more a cultural than a political one.

Because of this, politically active senior citizens demand institutionalised political representation of their interests. But the following examination of the political participation of senior citizens is an instructive example of the lasting political ambivalence in Germany concerning direct political participation.

Although senior citizens show a high turnout at elections, especially the age group 60–69 (Werner, 2003, p. 177), and a high share of members in the German parties SPD, CDU/CDS and PDS, the active political participation of older people is judged as too low (Naegele, 1999). In contrast to the USA, for example, the political representation of interests is weak. There is one political party for older people, 'The Greys', but its political success is extremely low. Naegele (*ibid.*, p. 132) sees this as a sign of the very heterogeneous interests of older people, of the lack of initiatives concerning exclusively older people, and of an ongoing strong bond with the established political parties.

Nevertheless, since the 1970s we find political attempts to improve the political participation of older people by communal senior citizen advisory boards. Today, more than a thousand boards have been established (Eifert, 2006). Each federal state has its own agency and the Federal Working Group of the States' Senior Citizens' Advisory Boards provides national representation. Until now the establishment of local senior citizens' advisory boards has been a voluntary accomplishment of the community; there is no legal claim for them in the community code. As a consequence, some communities refuse to establish a board, financial support for the boards varies widely, members of the boards have no legal claim to speak in local committees or to file a petition, and their integration into local policy making differs to a great extent (Frerichs and Kauss, 2001; Eifert, 2006).[4] The present refusal of most of the *Länder* to pass the legal integration of the boards into the community code, as much as the refusal of the Federal Ministry to guarantee a lasting form of financing for federal representation, are indicators of the existing ambivalence concerning direct political commitment in Germany.

The aims of the local boards are political representation of the interests of older people and involvement in consultation with local government. Other topics are the improvement of the local service structure and of the general quality of life in old age. From the beginning the senior citizens' advisory boards were politically and academically controversial. The critique aims at the insufficient democratic legitimacy of the boards, the insufficient political independence of political parties and associations, the social bias of their political commitment, the limited possibilities of political impact, the limited targets that the boards pursue, and last but not least their special status – as policy for senior citizens is a cross-sectional task, a specific representation of their interests is seen as counterproductive (Frerichs and Kauss, 2001, pp. 107ff).

However, the work of the boards can only be as good as the local conditions they are entitled to, and in most of the cases these conditions are limited (Eifert, 2006). Until now only one federal state, North Rhine-Westphalia, supports the local senior citizens' advisory boards with an expert employee, who acts as a consultant on the voluntary work. To improve the work of the boards the Commission of Enquiry 'The Future of Civic Activities' (Enquete-Kommission 'Zukunft des Bürgerschaftlichen Engagements', 2002, p. 217) recommended their nationwide establishment and a unique legal definition of their political rights.

The UK perspective

In the UK, the refocusing of the socio-political relationship between government (both local and national) and senior citizens has seen the establishment of a number of key initiatives. The year 2005 was designated as the Year of the Volunteer in an effort to encourage and broaden the involvement of individuals of all ages in their communities; to actively encourage 'civic engagement'. Yet this initiative aside, there is no clear evidence of heightened party-political activity among older people, nor are there political parties for older people that have any serious impact upon the political system just as in Germany. This is solely a direct result of the British electoral system, which does not encourage smaller, single-issue parties. However, similarly to Germany, older people in the UK tend to vote and participate in elections more than younger age groups.

At the local level, the formation of local strategic partnerships (LSPs) that facilitate inter-agency working and are responsible for formulating and implementing community-based strategies promoting and enhancing economic, social and environmental well-being is one main development. For example, LSPs provide an integrated approach to health care policy and the implementation of the National Service Framework (NSF) for Older People (Department of Health, 2001b). So the implementation of this policy is not generally instituted through specific newly created organisational structures as it is in Germany. Rather, it operates through already existing governmental agencies (both local and national) in partnership with the voluntary sector and NGOs. The most important example of this policy has been the strategy to encourage local participation in policy development and planning over recent years, enshrined within the 'Better Government for Older People' (BGOP) initiative.

The BGOP project started in 1998 and was consistent with the modernising agenda and new partnership approach of the Labour government (Lewis, 2005). It had support from central government and the direct involvement of a partnership of organisations (mainly NGOs) working for older people. These included Age Concern England, Help the Aged, Anchor Trust, Carnegie Third Age Programme and Warwick University. The aims of the programme were (Better Government for Older People, 2004):

- making a difference to the lives of older people as citizens;
- engaging older people in decision making;
- working in partnership to share experience and learning.

The programme was centred on 28 locally based two-year pilot projects organised around local authority areas throughout Great Britain and involving over 300 organisations. The role of these pilots was to test and develop innovative strategies by adoption of the central aims of the programme as outlined above. Thus it was a bold experiment to involve older people within the democratic model of involvement or participation. Details of these pilots and their contributions are summarised in the BGOP report *Making it Happen* (Better Government for Older People, 1999) which reported on the first year.

At the end of 2000 and with the final evaluation of the pilot projects undertaken by Warwick University, the project steering committee wrote a final report, *All Our Futures* (Better Government for Older People, 2000), making 28 recommendations in the following key areas:

- combatting age discrimination;
- engaging with older people;
- better meeting of older people's needs;
- promoting a strategic and joined-up approach.

Following these recommendations the UK government responded positively to each of them in its report *Building on Partnership* (Department for Work and Pensions, 2001) and called for the continuation of the programme through the establishment of a national partnership body to advise on policy issues affecting older people. The National Partnership Group (NPG) was established in 2002 to advise central government and included direct representation from an Older People's Action Group which had an important advisory and monitoring role in BGOP. BGOP continues its work through the NPG and via a number of 'strategic alliances' formed to both influence and be influenced by organisations whose desire is to ensure better services, better coordination and better joint working, leading to better governance. The BGOP initiative must be seen as constituting a sea-change in the nature of engagement with older people and their involvement in decision making and policy development. It has encouraged the wider engagement of older people in society and their empowerment, but more especially has placed the policy concerns of older people on the political agenda. As an example of this broader engagement, the Joseph Rowntree Foundation (a research-funding organisation) commissioned the Older People's Research Programme which was developed with and by ten older people who formed the steering group from 1999. Again, the National Service Framework for Older People has adopted a similar participative approach to decision making.

Trends and future prospects

This chapter has outlined a number of responses to encourage the participation of older people in our respective societies. Yet above all this, a fundamental and indeed central issue remains: how can a social and political consensus be built to ensure that the challenges of an ageing society are met equitably and to encourage the direct participation of all older people? The policy focus has hitherto often been constructed within an ageist framework in which older people tend to become pathologised, socially isolated and excluded.

If we review the present political efforts in both countries, we come to the following conclusions. In Germany the promotion of participation is focused on social integration via civic commitment. The role model of 'active and productive seniors' is an important attempt to change the social image of older people and prevent exclusion and isolation. But there are some general dangers concerning the strategy of reducing 'participation' to 'civic commitment'. First, these strategies mainly draw in already privileged seniors. Until now there are few efforts to reduce the inequality in social participation in old age, although economic polarisation in old age is growing (Frerichs and Naegele, 1999, p. 170; Bröscher *et al.*, 2000). Second, the promotion of civic commitment might change to a demand for civic commitment, the demand that senior citizens commit themselves to socially useful tasks. But the responsibility for solving the challenges of social and demographic shifts cannot be limited to the individual (older) citizen; it is a responsibility of the society as a whole (Aner, 2002, p. 39). Third, there is the underlying danger of a social division into 'good' (i.e., productive) seniors and 'bad' (dependent, unproductive) old people. The political efforts to foster a positive image of old age by stressing the abilities and the performance of older people should not lead to a situation where age-related losses and dependency in old age are devalued (Schmitt, 2004, p. 146). Besides this, the acceptance of civic commitment as a substitute for labour must be based on sufficient retirement pensions. Nevertheless, German pension policies as well as new labour market policies might have negative effects on financial security in old age, especially for older long-term unemployed people. Last but not least, the focus on civic commitment has a functional character – how can the potential and the abilities of older people be used? Policies to foster participation in the sense of involvement in decision making and user involvement have been less developed up to now.

Similar dangers also present themselves in the UK case too. Too often participative strategies have been responded to by those already engaged in their communities and not by the housebound and frail who as a result cannot be listened to or share their views: these are the 'lost voices'. Policy has got to start engaging with these individuals. The policy focus in the UK on greater participation has been different to the German model. As elaborated above, it has been very much an attempt to engage with people, to awaken a new 'civic renewal' among all citizens and develop better partnership working through the 'compact': a bottom-up approach rather than the previous Conservative administration's top-down model (Lewis, 2005).

It is timely, therefore, within the prevailing political and social climate, to focus upon the positive contributions that older people can and do make to society and upon a reconstitution of the meaning of old age. One vision lies within the concept of 'active ageing'. As Walker (2002a) describes, its genesis can be traced back to the 1960s and to the 'disengagement' thesis (cf. Cumming and Henry, 1961) and the promotion of 'successful ageing': the denial of the onset of old age (Havinghurst, 1954). A key mover in the newer structural interpretation of this and in the formulation of 'active ageing' has been the work of the World Health Organisation (WHO). It was they who summarised the meaning of 'active ageing' within the slogan, 'Years have been added to life; now we must add life to years'. This, as Walker (*ibid.*, p. 124) suggests, means that 'active ageing' is 'a general life-style strategy for the preservation of physical and mental health as people age, rather than just trying to make them work longer'.

It thus links quality of life with meaningful and productive employment, with mental and physical well-being. It could result in maximisation of the full potential and enhancement of the quality of life of all of society. It has the potential to move the policy focus away from pathologising older people as a distinct group, towards recognising that we are all ageing, constantly. As Walker (2002a) again suggests, 'active ageing' can provide a meaningful focus for policy development and has enormous social and economic potential for society at large.

A joint paper by the Association of Directors of Social Services and Local Government Association (ADSS, 2003) picks up on this theme and calls for a change in our perception of older people in UK (and indeed European) society, away from viewing them as dependants with a focus on acute care of the most frail. They present the provision of services as a triangle (see Figure 10.2) with acute care at the top of the service hierarchy. The radical change they suggest is for future provision of services to remodel itself so that there is a shift towards prevention and encouragement of community-based strategies: to invert the triangle (Figure 10.3). Older people need to be at the centre of such change.

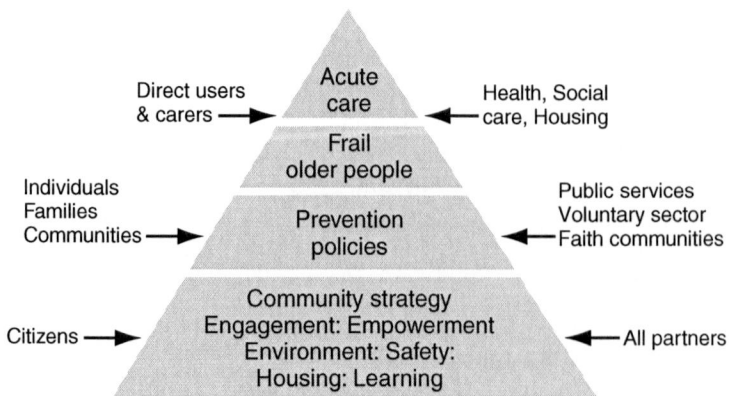

Source: ADSS (2003), p. 9.

Figure 10.2 Support for older people today

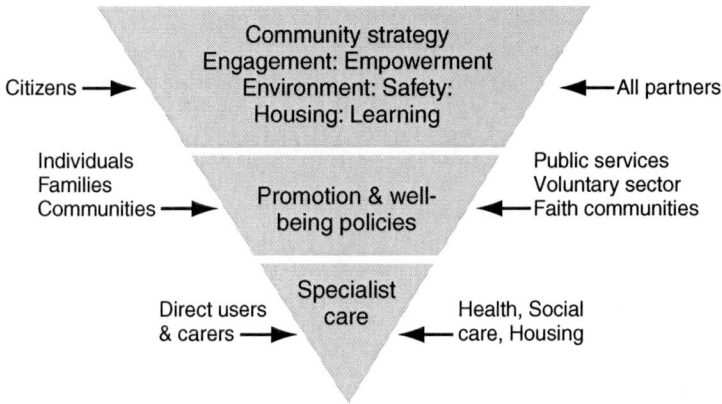

Source: ADSS (2003), p. 9.

Figure 10.3 Support for older people in the future

Notes

1. For example, the Social Ministry of Baden-Württemberg's 'Action Programme: Older Generation in the Centre' (Sozialministerium Baden-Württemberg, 2002), or the Ministry for Health, Social Affairs, Women and the Family of North Rhine-Westphalia's 'Age Designs the Future: Policy for Older People in North Rhine-Westphalia, Guidelines to 2010' (Ministerium für Gesundheit, Soziales, Frauen und Familie des Landes Nordrhein-Westfalen, 2004).
2. It is assumed, but until now not empirically proven, that there is an ethnic selectivity too (Olbermann, 2003, p. 70).
3. In the cooperatives more or less elderly people supply different social services for older and disabled people. In doing this they acquire their own claims for help if they become dependent themselves. Further programmes in Baden-Württemberg embrace the new priorities of building an age-unrelated, general infrastructure for civic commitment, with one main emphasis being on cross-generational commitment in the fields of home and stationary care.
4. North Rhine-Westphalia is the only federal state where the participation of the senior citizens' advisory boards in the local conferences for care is established in law (Eifert, 2006).

Conclusion

Gerhard Naegele and Alan Walker

The purpose of this book has been to provide an overview of key topics on ageing and social policy from an integrated Anglo-German perspective. Of vital importance to us in planning this publication was the *common* perspective of experts from the two countries. This approach to comparative research is without parallel in both fields, of ageing and social policy. It worked even better than we had anticipated and each chapter identifies both similarities and differences between the two countries and is able to reach specific conclusions and recommendations based on this comparison. There is no doubt that without the support of the Anglo-German-Foundation this innovative approach would not have been possible.

Together the individual contributions emphasise the complex problems that international and cross-cultural policy comparisons are faced with. Ultimately similarities as well as differences only become comprehensible if we acquire a good grasp of the development of attitudes, concepts and structures in the other country. The chapter authors have endeavoured hugely to do so and these efforts have paid dividends in terms of the quality of the topic case studies. Of course it was not possible to deal with all topics and policy fields that lend themselves to a comparison within the scope of this book and of the two-day conference that started the ball rolling.

The book deals with key policy fields of social policy with regard to ageing and older people on the one hand and, on the other, dimensions of their life circumstances of high social and political relevance. When analysing the differences between the countries in terms of concepts and measures, it is essential to pay special attention to the influence of different types of welfare regimes (Chapter 1). Indeed we can discern differences but also harmonisation between the regimes in the field of social policy for older people (Naegele and Walker, 2007).

Pensions

Pension policy and pension reforms are currently a very high priority for both countries. Ginn, Fachinger and Schmähl (Chapter 2) emphasise the fact that, so far, the differences dominate: the old-age security systems in Germany and in Great Britain are based on different social state conceptions. In Germany, the old-age security system resolves into numerous individual systems that are generally oriented to employment status. The dominant system, the statutory pension insurance, has been based on an earnings-related system of contributions and benefits since 1957 and thus is modelled on the insurance principle. The benefit level is relatively high, so that pensioners are less than average – in relation to the total population – affected by poverty. In contrast, Great Britain has a state pension system for all employees that generates a basic income, which, however, is not needs-based and, therefore, lies below the poverty line. This basic pension is complemented by an income-oriented state pension supplement. All in all, this does not lead to an appropriate level of old-age income, so that old-age poverty still represents a significant problem in the UK. The old-age poverty rate in Britain is twice as high as that in Germany.

Notwithstanding these differences a convergence of sorts can be observed. Germany has begun a pension reform policy that, to exaggerate a little, is moving towards the British model with some privatisation. Furthermore there is a growing call to change the financing over to the funding principle; the latter usually with reference to population ageing, although experts have repeatedly demonstrated that the funding principle is no less dependent on demography (if such a dependency exists at all). In addition, there has been a paradigm shift in the goal orientation: from a benefits to a contribution rate orientation. Against this background, the question of Ginn, Fachinger and Schmähl is justified: 'Why should private pensions be preferred to state pensions?'. This corresponds to the fact that the pension reforms carried out in Germany so far will in all likelihood lead to old-age poverty again; at any rate they will cause a hitherto absent skewed distribution of old-age incomes. Social inequality in old-age incomes – a characteristic of the UK today and of Germany tomorrow? And why this paradigm shift in German pension policy? An obvious answer might be: 'Political ideology has distorted and amplified the macroeconomic consequences of population ageing in order to legitimate anti-welfare policies' (Walker, 1990, p. 377).

Ginn, Fachinger and Schmähl conclude that, against this background, the development of the pension system in Britain and the current situation can, if anything, act as a deterrent for Germany. The transformation into a system that consists of a basic state pension and private and company supplementary pensions leads to a considerable amount of old-age poverty if the level of the basic pension is low. Moreover, the experience in Britain reveals which risks are associated with retirement arrangements organised by the private sector. This in turn points to the considerable importance of state control (or strict regulation) of private and company pension schemes as well as to the pertinence of low-risk forms of investment.

On the other hand, given the structural changes in the labour market and the reduction in jobs subject to social insurance contributions, the compulsory insurance for all employees in Britain proves to be advantageous. Such an employee insurance would lead to a reduction of the external effects of, among other things, the outsourcing by companies and of numerous economic policy measures, which have brought about an increase in the number of people without pension insurance in Germany. Conversely, the example of Germany shows the UK that a core pension system for a large part of the insured should have a benefit level above the poverty line, if old-age poverty is to be avoided.

Employment and training

In Germany and in the UK, the issue of the employment of older workers has moved up the policy agenda in recent years (Walker, 1997). After decades of early retirement, concerns about the sustainability of public pension systems and future labour shortages have resulted in a new policy consensus around the need to reintegrate older workers into the labour force. Currently, policies are emerging which aim at extending working lives, at closing early-retirement pathways and at making continued employment more attractive, and, in addition, at educating employers and encouraging them to recruit older workers. Yet the degree of progress should not be overemphasised. Considering active labour market policies in detail and reviewing the evidence, Frerichs and Taylor (Chapter 3) conclude that, while a recent shift towards strategies and measures for older unemployed people has taken place, they remain under-represented in general labour market measures such as the promotion of vocational training, self-employment and job placement activities. In particular, strategies and measures to facilitate lifelong learning and to provide training for all age groups are underdeveloped. Even where measures have been implemented, they have concentrated on more advantaged and easier-to-place older unemployed people, with at-risk groups such as those with low skill levels or disabilities remaining under-represented.

The primary deficit concerning active labour market policies for the older unemployed in Germany is the lack of specific targeting of this group both in active job placement and in training. In the UK, the scope of active measures is still rather limited both with regard to the kind of measures – New Deal 50 Plus /New Deal 25 Plus – and the level and duration of funding. Moreover, participation by the most severely disadvantaged older people (e.g., those with disabilities) is low.

Furthermore, there is still a distinct mix of underlying frameworks and structures for policy measures in both countries. The pressures of labour force and population ageing are tackled on the basis of the respective welfare regimes, economic structures and normative values. In the UK, despite a more socially inclusive stance recently, funding of job creation and a broad application of training measures have not taken place to any significant extent so far, and the low-intervention character of labour market policies has been maintained. In Germany, in the wake of recent labour market reforms, a shift in paradigm towards a more activating approach to job placement has taken place, which builds upon an increased responsibility of the unemployed for job search activities and curtails unemployment benefits. However, a wide array of active labour market measures regulated at the federal level is still in place.

Despite these deficiencies and differences, mutual policy learning can take place. Because a close coordination of public and workplace policies is needed as well as better governmental coordination, an exchange of experiences between governmental actors in Germany and the UK could be very fruitful. Furthermore, concerning job placement policy, Germany could borrow from the UK by adopting its system of credit points for the hard-to-place and by increasing staff resources in job centres. To foster training measures for older workers and to increase their employability, the UK could adopt the short-term training measures for the unemployed, job rotation and promotion of vocational training for older workers in small and medium-sized enterprises (SMEs) currently in force in Germany. As well as these concrete recommendations towards improving the effectiveness of active labour market policies, we have to stress the need for both countries to establish an integrated and life-course-oriented labour policy approach in conjunction with pension and taxation policies, policies for promoting health and education and policies to enhance the work–life balance.

Older consumers

The surprising conclusion of Potratz, Gross and Hilbert (Chapter 4) on the 'silver economy' is that differences are larger within the respective national age cohorts than between the countries. In both countries there is growing wealth and hedonism concerning consumption and lifestyle and a supply gap with respect to age-specific demand. The explanation is to be found in explicit welfare policies after World War II and the long-lasting growth periods. Age is no longer mainly associated with poverty, poor health and economic irrelevance. Rather the opposite holds true: increasing income has fostered the development of stable consumption patterns, expectations and perceptions, which now, after retirement, are perpetuated for as long as possible. Travelling is a good example: people who have experienced the fascination of travelling during their 'active' times will go on travelling, with more or less different demands concerning quality and convenience, perhaps with a reduced range of destinations. This can be observed particularly in Germany, where 'silver travellers' have become the most courted group of customers of travel offices. Similarly, health, which ranks high on retirees' agendas in both countries, generates a growing and increasingly sophisticated supply of health products and activities.

Nevertheless income remains the core variable which decides about access to and use of offers for greater quality of life in old age. Although levels have gone up considerably in both countries, differences within age cohorts are considerable, too. What comes to bear here is not only former professional status and related income, but also other sources of income such as real estate, rents or additional private pensions. Differences also cluster regionally: while in Germany there is a clear East–West divide due to the developments since 1990, in the UK we find the better-off in the South in and around London rather than in the North.

The centrepiece of an independent life in old age, however, is housing, and it is probably here that (relative) differences are the largest. Social and professional status, income, wealth, family relations add up to a social and economic 'startup capital' which reinforces pre-existing differences. The inappropriateness of much of each nation's housing stock, which was built with little regard for the needs of older people, is a major structural problem, and indicates market deficiencies. Here the chapter goes further and builds a bridge to household services which link 'housing' with 'living' and presents data on which services are demanded and which services seniors would be prepared to pay for, if they were on offer.

This is the central lesson of this contribution: not to look at single products and services but at how they link up to provide more quality of life in old age. It is by initiating such demand-multiplier processes that a market is established, industries are constituted and economic growth and employment are triggered.

Gender

Gender aspects, for demographic reasons alone, have always played an important role in the gerontological research of both countries but less so in social policy towards older people. In Germany typical social risks of (older) women (such as the loss of social protection in old age in the case of divorce) were for a long time either not recognised as a policy issue or were neglected in the face of the general objectives of family policy (which was, *de facto*, a 'marriage policy'). As shown by Eyers and Backes (Chapter 5), women are by far the more active in the everyday lives of older people in both countries as far as the organisation and maintenance of social networks are concerned; and the traditionally gender-typical division of labour is widespread, especially in old age, in both countries. Therefore, older women have comparatively more 'social capital' (Bourdieu, 1983) in both the UK and in Germany, *inter alia* an expression of gender-specific parenting styles and education conceptions in the past in both countries. From a socio-political point of view, this is all the more significant, as existing social networks influence quality of life and non-material security in old age.

All of this holds good equally in both countries. As far as this life circumstance factor is concerned, there are (not surprisingly, as the gender-typical division of labour is still interculturally a widespread social phenomenon) more similarities than differences between the countries. However, it is expected, and again this holds true for both countries, that as Eyers and Backes write, 'Gradually gender equality and conformity in European policies is diminishing the differences between gender in the UK and Germany'. This applies especially to the baby-boomer cohorts; thus, once more, there are more similarities than differences!

Ethnic minorities

As opposed to the UK, Germany has only limited experience in dealing with older migrants. In Germany, the phenomenon of 'growing old in a second homeland' only stems from the beginning of the

1990s and moreover centres on very few regions, in particular on the industrial centres, to which the 'guest workers' of the first generation moved, where they worked and, unexpectedly for many, remained. In this respect, the comparison between the UK and Germany undertaken by Gerling (Chapter 6) rests upon completely different experiences and historical backgrounds.

The life circumstances and the quality of life of older migrants in the UK are influenced decisively by the traditional special responsibility of the communities, which often systematically incorporate older migrants and provide support. In addition, the anti-discrimination legislation is of crucial relevance. In Germany there are no comparable traditions. The charities that are in fact responsible for addressing the social problems of migrants were traditionally occupied with questions of the right of residence and employment law, but not with problems of ageing in a second homeland. For this reason alone they reach migrants less and less. Furthermore, minority ethnic social services with a culture-specific background are still largely in a conceptualisation or, at best, in a startup phase. This affords opportunities for looking to the UK for examples of good practice. All the more surprising therefore that this has not yet been done systematically in Germany. Gerling's contribution, and the extensive case studies in Leeds and in Dortmund it is based on, offer numerous concrete starting points, in particular at the communal level. Even if Gerling critically observes that social services for older migrants only play a subordinate role in the overall context of social services for minority ethnic groups in the UK, the experience gained is likely to be instructive for Germany. Important too (and worth copying in Germany) is the fact that the established elder organisations in the UK such as Age Concern and Help the Aged also attend to this issue.

Intergenerational relations

In both countries, the relations between the generations are considered an important indicator for social cohesion. As already shown by many earlier studies, there can be no question, at least at the micro level, of a cancellation of the 'small generation contract' or even of a collapse of family solidarity in Germany or in the UK. As Reichert and Phillips (Chapter 7) demonstrate, in both countries inner-familial solidarity is practised to a high degree in the case of illness or the need of care. The generations assist each other in financial respects too, even if in the UK the cash flows often go more in the direction from the young to the old (in Germany, this is the other way around). This is an expression of the more unfavourable (favourable) economic situation of older people in the UK (in Germany). Despite their different welfare regimes this corresponds to a surprisingly high agreement among seniors in both countries to the effect that the right strategy for the problem of older people in need of care and help is to make them the joint responsibility of state and family. However, whether this will continue to hold good in future is still an open question in both countries, given further changes in the family and in social protection. Thus there are already indications in both countries that intra-familial solidarity is threatening to break down, if the problem of the reconciliation of work and care has not been solved to the satisfaction of those care givers who have a job. In both countries this is a challenge for the welfare state.

Against this background, the question is increasingly pressing whether and, if applicable, what repercussions the withdrawal of social security benefits and the growing trend towards the privatisation of social risks will have. If it is true that a welfare state framework has a beneficial, supportive influence on intra-familial help patterns (the 'big generation contract' as the material and/or ideological basis for the 'small generation contract'), then the potential damage created by threatening developments cannot be excluded. In the UK as well as in Germany we identify 'secondary arenas', in which attempts are made to take countermeasures. For instance, within the scope of the 'work–life balance' initiative or the anti-discrimination campaign in Great Britain, which Reichert and Phillips refer to, or in the context of the current productivity debate in Germany, which Maltby and Rohleder (Chapter 10) describe. But 'secondary arenas' are what they are. At present, the main threat to generational solidarity comes from the 'main arenas' of social protection policy (Walker, 1997). Thus we concur with Reichert's and Phillips' conclusion: 'Similar problems appear in both countries and both countries have to answer the challenge. But Britain and Germany need more: they need a policy which is sensitive to generational issues in general'.

Health and health care

Health is a critical determinant of quality of life in old age (Walker, 2005; Walker and Hagan Hennessey, 2005). In Chapter 1 we outlined the different welfare regime paths trodden by Germany and the UK and, as Pfaff illustrates in Chapter 8, these are reflected in their contrasting health and social care systems. Despite being co-members of the EU and OECD the two countries have very different health care systems and, in practice, different ideas about key aspects of health care such as need, quality and who should be covered. In sum, whereas the UK chose in 1948 a Beveridge-style model of universal coverage with minimum quality levels, it was not until 2007 that Germany introduced universal coverage (to be implemented by 2009).

With its roots stretching back to the 1870s the German health care system aimed for quality rather than quantity and only gradually extended its coverage to more groups in the population (including older people in 1941). The two systems also contrast in the level of private provision they include, with the UK still adhering largely to the idea of health as a public good. They contrast further in the levels of expenditure devoted to health care as a proportion of national income. Germany allocates a higher proportion of GDP to health care than the UK (10.6 per cent compared with 8.1 per cent). As a result levels of provision are higher in Germany. This has meant that the endemic problem of NHS waiting-lists has not been mirrored in Germany. Despite this difference, however, major health outcomes appear to be similar.

With regard to the impact of the two systems on older people, Pfaff shows some convergence. For example, Germany is in the process of universalising the coverage of its system while the UK is attempting to raise the quality of its NHS. Moreover the two countries' health care systems face similar challenges, including managing the tradeoff between economic efficiency and service quality (there is a danger here that hospitals seeking to reduce their financial risks will attempt to avoid treating some older people). The neglect of older people suffering from depression is a longstanding problem in both countries. Both Britain and Germany are trying to use information and communications technologies (ICTs) to improve patient information though, because of its history of underfunding, the NHS starts from a weaker position than the German health service.

The differences in system design and resourcing have a disproportionate impact on older people because they are the main users of health services. Both systems can be expected to become better adjusted to the changing demographic structure and, as Pfaff notes, become more effectively geared to the needs of older people. It is unlikely, however, that older people will escape from the consequences of future economic constraints on the health care systems of the two countries.

Social care

As Glendenning and Igl (Chapter 9) emphasise, in the field of social care and care policy the differences between Germany and the UK dominate despite more or less comparable prevalence rates, care structures (in each case a preponderance of home care by close relatives) and expenditure rates (as measured by proportion of GDP). The reasons for the differences can be found in the specific historically influenced approaches of German and British policy makers to the problem of care, relating to organisation and responsibility, to questions of funding and traditionally prevailing cultural patterns of care. However, it is difficult to come to a final evaluation because both systems have specific advantages and disadvantages. Ultimately, the user perspective has to serve as an assessment framework, and in this respect the traditional community proximity of care responsibility and organisation in Britain seems to have an advantage over the situation in Germany.

In practice, the specific differences lie in the core competencies (decentralisation of responsibility in the UK), the funding structures (multiple competencies in the UK), the right to benefits (universality in Germany vs selectivity in the UK), the access criteria for benefits (virtually equity in Germany vs varying criteria between countries and sectors in the UK), the choices between different kinds of benefits (more pronounced in Germany than in the UK) as well as, last but not least, in the treatment of familial care givers. Especially in this respect, the British situation is clearly superior, as in Britain the main care givers have a direct (independent) right to money and other benefits, not as in Germany only an imparted (quasi-derived) right via the persons in need of care; a fact indicating that the role of the private care giver enjoys public recognition and protection in the UK.

The responsibility of local communities for social and health services in the UK, which can also be found in the field of care as regards the responsibility for planning, coordination and supervision tasks, is likewise in direct contrast to the German tradition. On the individual level, the British path of care management continues to be convincing from a German point of view. Germany is just painstakingly and overcautiously (by means of pilot projects) beginning to follow this path (which is being downgraded again in the UK in favour of a strengthening of the direct purchasing role of the persons in need of care, as in Germany). Furthermore, the longer and more pronounced British tradition of safeguarding the quality of care must be emphasised; this is, *inter alia*, also an expression of a traditionally stronger user-orientation and participation in the social services field. If Germany were to match the UK in this respect, then according to Glendinning and Igl, 'this could represent a profound convergence between two currently very different approaches to long-term care'.

Participation

In the light of demographic change, the question as to how the societal and political participation of older people can be effectively improved is on the agenda in Germany and the UK, as Maltby and Rohleder explain in Chapter 10. In this respect, anti-ageism movements play a role in both countries. Nevertheless, in practice the two countries take different stances: Germany primarily seeks to improve participation via the role model of 'active and productive seniors' and, following a top-down approach, by more or less bluntly calling for active participation in society. However, the risk of exclusion of and of discrimination against non-active, non-productive older people (older migrants, sick persons and those in need of care, the very old and frail) is thus partly predetermined. The *Fünfter Bericht zur Lage der älteren Generation in der Bundesrepublik Deutschland* (Fifth National Report on the Situation of the Older Generations in Germany) (BMFSFJ, 2006a), while acknowledging this risk, is also of no assistance in this respect – at least as far as concrete strategies are concerned.

In contrast, the British model is at least a step ahead. The policy makers are aware of the risk of excluding vulnerable older people and focus on a bottom-up approach, which attempts, in the words of Maltby and Rohleder, to achieve 'a new "civic renewal" among all citizens'. This strategy is integrated into the overarching concept of 'active ageing' that, among other things, explicitly aims at promoting the social participation of *all* groups of older people, including the socially weaker and disadvantaged groups. For Germany, a conceptualisation of the social and political participation, or social exclusion, of older people that includes all social groups in such a way, is still pending.

Closing remarks

This comparative endeavour has revealed systemic and operational contrasts and coincidences. The intention has been not to suggest that one is better than the other in any respect, but to explore the extent of the differences that reflect the original welfare state designs and the subsequent, often dependent, paths that the two major European welfare systems have taken. It has demonstrated that, regardless of differences, there are important experiences to be shared in the social policies on ageing between the two countries. This may, on the one hand, prevent policy errors being repeated and, on the other, facilitate the replication of good practices in responding to societal ageing. It is evident too that the stories of social policy and ageing in these two archetypal European welfare states contain much that should be of interest to the rest of the EU and, especially, its newest members. The evidence presented in this book provides ample demonstration of both the centrality of ageing to social policy, in operational terms if not yet, sadly, in its academic pursuit and, also, the extent of the 'structural lag' experienced by both Germany and the UK in adjusting to unprecedented demographic change.

Bibliography

ADSS (2003) *All Our Tomorrows: Inverting the Triangle of Care*. London: Association of Directors of Social Services / Local Government Association.

Age Concern England (2001a) *Ethnic Elders: Access, Equality*. Report of a Conference held in London, 12 November 2001.

Age Concern England (2001b) *Ethnic Elders: Access, Equality: The Impact of Government Policy for Black and Minority Ethnic Elders*. London: Age Concern England.

Age Concern England (2003a) *BME Elders Forum Newsletter*, 1, May.

Age Concern England (2003b) *BME Elders Forum Newsletter*, 3, November.

Age Concern England (2004a) *BME Elders Forum Newsletter*, 4, March.

Age Concern England (2004b) *BME Elders Forum Newsletter*, 5, June.

Age Concern England (2004c) *BME Elders Forum Newsletter*, 6, October.

Age Concern England (2005a) *BME Elders Forum Newsletter*, 7, February.

AGE Concern England (2005b) *BME Elders Forum Newsletter*, 8, May.

AGE Concern England (2005c) *BME Elders Forum Newsletter*, 9, September.

Age Concern England and Commission for Racial Equality (1995) *Age and Race: Double Discrimination: Life in Britain Today for Ethnic Minority Elders*. London: Age Concern England.

Age Concern England (2005d) *Policy Position Papers: Employment*. London: Age Concern England.

Age Concern England and Ethnic Minorities Steering Group (EMSG) (1999) *Ageing Matters: Ethnic Concerns*. Report from Naina Patel. London: PRIAE.

Age Concern England and Housing Corporation (2005) *Aims for Black and Minority Ethnic Elders*. London: Age Concern England.

Age Exchange (ed.) (2004) *Mapping Memories – Reminiscence Work with Ethnic Minority Elders*. London: Age Exchange.

Agulnik, P. and Le Grand, J. (1998) 'Tax Relief and Partnership Pensions', *Fiscal Studies*, 19(4), pp. 403–28.

Ahmad, R. (2002) 'The Older or Ageing Consumers in the UK: Are They Really That Different?', *International Journal of Market Research*, 44, Quarter 3.

Alber, J. (1996) 'The Debate about Long-term Care Reform in Germany', in OECD, *Caring for Frail Elderly People*, Policy Studies 19, Paris: OECD, pp. 261–78.

All Party Parliamentary Group on Ageing and Older People (2006) *Older Workers and Incapacity Benefit Reform*. Retrieved from www.ageconcern.org. uk/AgeConcern/Documents/Older_Workers_and_Incapacity_Benefit_Reform_Inquiry.pdf.

Allen, I., Hogg, D. and Peace, S. (1992) *Elderly People: Choice, Participation and Satisfaction*. London: PSI.

Allianz Group (2004) *Demography, Savings and Yields: Long-term Outlook*. Economic Research Working Paper No. 21.

Alterssicherung in Deutschland (2003) (ASID '03) *Zusammenfassung wichtiger Untersuchungsergebnisse*. Untersuchung im Auftrag des Bundesministeriums für Gesundheit und Soziale Sicherung (durchgeführt von TNS Infratest

Sozialforschung) München 2005 (English summary version: Old-age Pension Schemes in Germany 2003 (ASID '03) Summary of Survey Results. Research project commissioned by the Federal Ministry of Health and Social Affairs. Munich, 2005).

Aner, K. (2002) 'Das freiwillige Engagement älterer Menschen Ambivalenzen einer gesellschaftlichen Debatte'. In Karl, F. and Aner, K. (eds) *Die 'neuen' Alten revisited*, Kassel: Kassel University Press, pp. 39–102.

Aner, K., Karl, F. and Rosenmayr, L. (eds) (2007) *Die neuen Alten – Retter des Sozialen?* Wiesbaden: VS Verlag für Sozialwissenschaften.

Arber, S., Davidson, K. and Ginn J. (2003a) 'Changing Approaches to Gender and Later Life' in Arber, S., Davidson, K. and Ginn J. (eds) *Gender and Ageing: Changing Roles and Relationships*, Maidenhead: Open University Press, pp. 1–14.

Arber, S., Gilbert, G. N. and Evandrou, M. (1988) 'Gender, Household Composition and Receipt of Domiciliary Services by Elderly Disabled People', *Journal of Social Policy*, 17, pp. 153–75.

Arber, S. and Gilbert, G. N. (1989) 'Men: The Forgotten Carers', *Sociology*, 23(1), pp. 111–228.

Arber, S. and Ginn, J. (1991) *Gender and Later Life: A Sociological Analysis of Resources and Constraints*. London, Newbury Park, CA and New Delhi: Sage.

Arber, S. and Ginn, J. (2004) 'Ageing and Gender: Diversity and Change', lead article in *Social Trends 2004*, No. 34, London: HMSO, pp. 1–14.

Arber, S., Price, D., Davidson, K. and Perrin, K. (2003) 'Re-examining Gender and Marital Status: Material Wellbeing and Social Involvement' in Arber, S., Davidson, K. and Ginn J. (eds) *Gender and Ageing: Changing Roles and Relationships*, Maidenhead: Open University Press, pp. 148–67.

Askham, J., Henshaw, L. and Tarpey, M. (1995) *Social and Health Authority Services for Elders People from Black and Minority Ethnic Communities*. London: HMSO.

Atkin, K. and Rollings, J. (1994) *Community Care in a Multi-Racial Britain: A Critical Review of the Literature*. London: HMSO.

Atkinson, J. and Dewson, S. (2001) *Evaluation of the New Deal 50 Plus: Research with Individuals, Wave 1*. Sheffield: Employment Service.

Atkinson, J., Kodz, J., Dewson, S. and Eccles, J. (2000) *Evaluation of New Deal 50 Plus: Qualitative Evidence from Clients, First Phase*. Sheffield: Employment Service.

Attias-Donfut, C. and Arber, S. (2000) 'Equity and Solidarity across the Generations', in Attias-Donfut, C. and Arber, S. (eds), *The Myth of Generational Conflict: The Family and State in Ageing Societies*, London: Routledge.

Audit Commission (2004a) *Older People – A Changing Approach*. London: Audit Commission. Retrieved from www.audit-commission.gov.uk/olderpeople/ olderpeoplereports.asp.

Audit Commission (2004b) *Assistive Technology*. Retrieved from www.audit-commission.gov.uk/Products/NATIONAL-REPORT/BB070AC2-A23A-4478-BD69-4C19BE942722/NationalReport_FINAL.pdf.

Augurzky, B. and Neumann, U. (2005) *Volkswirtschaftliche Kosten der Nichtbeachtung ökonomischer Ressourcen und Stärken älterer Menschen*. Regionalwirtschaftliche und fiskalische Effekte einer Förderung der Seniorenwirtschaft in Nordrhein-Westfalen. Essen: Expertise im Auftrag des IAT.

266 Bibliography

Bäcker, G., Bispinck, R., Hofemann,K. and Naegele, G. (2000) *Sozialpolitik und soziale Lage in Deutschland, Bd. 2: Gesundheit und Gesundheitssystem, Familie, Alter, Soziale Dienste.* Wiesbaden: Westdeutscher.

Backes, G. M. (1983) *Frauen im Alter* (2. Auflage) Bielefeld: AJZ (1. Auflage 1981).

Backes, G. M. (1993) 'Frauenerwerbslosigkeit und Alter(n)', in Mohr, G. (ed.) *Ausgezählt: Theoretische und empirische Beiträge zur Psychologie der Frauenerwerbslosigkeit.* Bremen: Deutscher Studien.

Backes, G. M. (2002) ' "Geschlecht und Alter(n)" als künftiges Thema der Alter(n)ssoziologie', in Backes, G. M. and Clemens, W. (eds) *Zukunft der Soziologie des Alter(n)s*, Opladen: Leske und Budrich, pp. 111–48.

Backes, G. M. (2003) 'Frauen – Lebenslagen – Alter(n) in den neuen und alten Bundesländern', in Reichert, M., Maly-Lukas, N. and Schönknecht, C. (eds) *Älter werdende und ältere Frauen heute: Zur Vielfalt ihrer Lebenssituationen*, Wiesbaden: Westdeutscher, pp. 13–34.

Backes, G. M. (2004) 'Alter(n) Ein kaum entdecktes Arbeitsfeld der Frauen- und Geschlechterforschung', in Becker, R. and Kortendiek, B (eds) *Handbuch Frauen- und Geschlechterforschung: Theorie, Methoden, Empirie*, Wiesbaden: VS Verlag für Sozialwissenschaften, pp. 395–401.

Backes, G. M. (2006) 'Widersprüche und Ambivalenzen ehrenamtlicher und freiwilliger Arbeit im Alter', in Schroeter, K. and Zängl, P. (eds) *Altern und bürgerschaftliches Engagement: Aspekte der Vergemeinschaftung und Vergesellschaftung in der Lebensphase Alter.* Wiesbaden: VS Verlag für Sozialwissenschaften, pp. 63–94.

Backes, G. M., Amrhein, L., Lasch, V. and Reimann, K. (2006) 'Gendered Life Course and Ageing – Implications on "Lebenslagen" of Ageing Women and Men', in Backes, G. M., Lasch, V. and Reimann, K. (eds) *Gender, Health and Ageing.* Wiesbaden: VS Verlag für Sozialwissenschaften, pp. 29–56.

Bäckes, G., Bispinck, R., Hofemann, K. and Naegele, G. (2000) *Sozialpolitik und soziale Lage in Deutschland, Bd. 2: Gesundheit und Gesundheitssystem, Familie, Alter, Soziale Dienste.* Wiesbaden: Westdeutscher.

Balchin, S. and Bullen, C. (2005) *The Pensioners' Income Series 2003/4, National Statistics.* Retrieved from www.dwp.gov.uk/asd/asd6/pi_internet_april05.pdf.

Baltes, M. (1987) 'Erfolgreiches Altern als Ausdruck von Verhaltenskompetenz und Umweltqualität', in Niemitz, C. (ed.) *Erbe und Umwelt: Zur Natur von Anlage und Selbstbestimmung des Menschen.* Frankfurt am Main: Suhrkamp, pp. 353–76.

Baltes, P. B. (2005) 'Zukunft ist Alter', paper presented at Schauspielhaus Zürich, 24 April.

Baltes, P. and Mayer, K. (eds) (1999) *The Berlin Aging Study: Aging from 70 to 100.* Cambridge and New York: Cambridge University Press.

Baltes, M. M. and Montada, L. (eds) (1996) *Produktives Leben im Alter.* Frankfurt am Main: Campus.

Banister, D. and Bowling, A. (2004) 'Quality of Life for the Elderly: The Transport Dimension', *Transport Policy*, 11, pp. 105–15.

Banks, J., Emmerson, C., Oldfield, Z. and Tetlow, G. (2005) *Prepared for Retirement? The Adequacy and Distribution of Retirement Resources in England.* London: Institute for Fiscal Studies.

Baringhorst, S. (1993) 'Multikulturalismus und Anti Diskriminierungspolitik in Großbritannien', in Robertson-Wensauer, C. Y. (ed.) *Multikulturalität – Interkulturalität? Probleme und Perspektiven der multikulturellen Gesellschaft.* Baden-Baden: Nomos, pp. 193–211.

Barnes, M. (1999) 'Users as Citizens: Collective Action and the Local Governance of Welfare', *Social Policy and Administration*, 33(1), pp. 73 –90.

Barnes, M. (2005) 'The Same Old Process? Older People Participation and Deliberation', *Ageing and Society*, 25(2), pp. 245–59.

Bauld, L., Chesterman, J., Davies, B., Judge, K. and Mangalore, R. (2000) *Caring for Older People: An Assessment of Community Care in the 1990s*. Aldershot: Ashgate.

Baur, R., Czock, H., Scheuerl, A. and Schirowski, U. (1997) *Gerontologische Untersuchung zur motivationalen und institutionellen Förderung nachberuflicher Tätigkeitsfelder: Die Aktion 55*. Schriftenreihe des Bundesministeriums für Familie, Senioren, Frauen und Jugend (BMFSFJ), Bd.130.2. Stuttgart, Berlin and Cologne.

BBC News (2006) 'Older Staff "Dumped" as Law Looms', 10 September. Retrieved from http://news.bbc.co.uk/2/hi/uk_news/5333100.stm.

BBC News (2007) 'Equality Law "Should Go Further" ', 12 June. Retrieved from http://news.bbc.co.uk/2/hi/uk_news/politics/6742955.stm.

Beck, U. (1986) *Risikogesellschaft*. Frankfurt an Main: Suhrkamp.

Beresford, P. (1992) 'Researching Citizen Involvement: A Collaborative or Colonising Exercise?', in Barnes, M. and Wistow, G. (eds) *Researching User Involvement*, Leeds: Nuffield Institute for Health, University of Leeds, pp. 16–32.

Beresford, P. (1997) 'The Last Social Division? Revisiting the Relationship Between Social Policy, Its Producers and Consumers', in May, M., Brunsdon, E. and Craig, G. (eds) *Social Policy Review 9*, London: SPA, pp. 203–26.

Better Government for Older People (BGOP) (1999) *Making it Happen*. Wolverhampton: BGOP.

Better Government for Older People (BGOP) (2000) *All our Futures: The Report of the Better Government for Older People Steering Committee*. Wolverhampton: BGOP.

Better Government for Older People (BGOP) (2004) www.bgop.org.uk/home.aspx, accessed August 2004.

Beveridge, W. (1942) *Social and Allied Services* (The Beveridge Report). Report presented to Parliament by Command of His Majesty, Cmnd 6404. London: HMSO.

Bhalla, A. and Blakemore, K. (1981) *Elders of the Minority Ethnic Groups*. Nottingham: Russell.

Bhatnagar, K. and Frank, J. (1997) 'Psychiatric Disorders in Elderly from the Indian Subcontinent Living in Bradford', *International Journal of Geriatric Psychiatry*, 12, pp. 907–12.

Biggs, S. (2004) 'New Ageism: Age Imperialism, Personal Experience and Ageing Policy' in Daatland, S. O. and Biggs, S. (eds) *Ageing and Diversity: Multiple Pathways and Cultural Migrations*, Bristol: Policy Press, pp. 95–106.

Blakemore, K. and Boneham, M. (1994) *Age, Race and Ethnicity: A Comparative Approach*. Buckingham: Open University Press.

Blinkert, B. and Klie, T. (2000) 'Pflegekulturelle Orientierungen und soziale Milieus: Ergebnisse einer Untersuchung über die sozial-strukturelle Verankerung von Solidarität', *Sozialer Fortschritt*, 10, pp. 237–45.

BME Elders Forum (2005) *Report of the Black and Minority Ethnic (BME) Elders Forum*. London.

Bond, J. and Corner, L. (2004) *Quality of Life and Older People*. Maidenhead: Open University Press.

Börsch-Supan, A. (2005) 'Economic Implications of Demographic Change'. Manuscript, Mannheim and Cambridge, MA.

Bourdieu, P. (1983) 'Ökonomisches Kapital, kulturelles Kapital, soziales Kapital', in Kreckel, R. (ed.) *Soziale Ungleichheiten*, Göttingen: Soziale Welt, Sonderband 2, pp. 183–98.

Braun, J. (2003) 'Förderung des bürgerschaftlichen Engagements auf Länderebene in Enquete Kommission "Zukunft des Bürgerschaftlichen Engagements" ', in Deutscher Bundestag (ed.) *Politik des bürgerschaftlichen Engagements in den Bundesländern*, Opladen, pp. 109–53.

Braun, J. and Bischoff, S. (1999) *Bürgerschaftliches Engagement älterer Menschen: Motive und Aktivitäten. Engagementförderung in Kommunen – Paradigmenwechsel in der offenen Altenarbeit*. Schriftenreihe des Bundesministeriums für Familie, Senioren, Frauen und Jugend (BMFSFJ), Bd. 184. Stuttgart, Berlin and Cologne.

Braun, J., Burmeister, J. and Engels, D. (2004) *SeniorTrainerin: Neue Verantwortungsrollen und Engagement in Kommunen*. Bundesmodellprogramm 'Erfahrungswissen für Initiativen' – Bericht zur ersten Programmphase, ISAB – Berichte aus Forschung und Praxis, Nr. 84. Leipzig.

Braun, J. and Emons, G. (2000) *Seniorenbüro: Beispiel für eine neue Altenarbeit in der Kommune: Nutzen – Einrichtung – Finanzierung*. Eine Veröffentlichung der Bundesarbeitsgemeinschaft Seniorenbüros. Stuttgart, Marburg and Erfurt.

Brendgens, U. and Braun, J. (2000) 'Freiwilliges Engagement älterer Menschen', in Picot, S. (ed.) *Freiwilliges Engagement in Deutschland: Ergebnisse der Repräsentativerhebung zu Ehrenamt, Freiwilligenarbeit und bürgerschaftlichem Engagement, Bd. 3: Frauen und Männer, Jugend, Senioren, Sport*. Schriftenreihe des Bundesministeriums für Familie, Senioren, Frauen und Jugend (BMFSFJ), Bd. 194.3. Stuttgart, Berlin and Cologne, pp. 209–301.

Breyer, F. and Kliemt, H. (1994) 'Lebensverlängernde medizinische Leistungen als Clubgüter?', in K. Homann, K (ed.) *Wirtschaftsethische Perspektiven I*. Schriftenreihe des Vereins für Socialpolitik, Neue Folge, Bd. 228/I, Berlin: Duncker and Humboldt, pp. 131–58.

Breyer, F. and Schultheiss, C. (2002) ' "Alter" als Kriterium bei der Rationierung von Gesundheitsleistungen: Eine ethisch-ökonomische Analyse', in Gutmann, T. and Schmidt, V. H. (eds) *Rationierung und Allokation im Gesundheitswesen*, Weilerswist: Velbrück Wissenschaft, pp. 121–53.

Breyer, F., Zweifel, P. and Kifmann, M. (2005), *Gesundheitsökonomik*, 5th ed. Berlin, Heidelberg and New York: Springer.

British Medical Association (BMA), Olbermann, E. and Dietzel-Papakyriakou, M. (1995) *Entwicklung von Konzepten und Handlungsstrategien für die Versorgung älterwerdender und älterer Ausländer: Abschlußbericht der wissenschaftlichen Begleitung und Beratung*. Bonn: Eigenverlag.

Broad, B. (ed.) (2000) *Kinship Care: the Placement Choice for Children and Young People*. Lyme Regis: Russell House.

Brookes, R., Regan, S. and Robinson, P. (2002) *A New Contract for Retirement*. London: Institute for Public Policy Research.

Bröscher, P., Naegele, G. and Rohleder, C. (2000) 'Freie Zeit im Alter als gesellschaftliche Gestaltungsaufgabe', *Aus Politik und Zeitgeschichte*, B35–36, pp. 30–8.

Buck, H. and Dworschak, B. (eds) (2003) *Ageing and Work in Europe: Strategies at Company Level and Public Policies In Selected European Countries*. Stuttgart: Fraunhofer IRB.

Bundesanstalt für Arbeit (BA) (2002) *Eingliederungsbilanz 2001*. Nürnberg: Bundesergebnisse.

Bundesanstalt für Arbeit (BA) (2003) *Eingliederungsbilanz 2002*. Nürnberg: Bundesergebnisse.

Bundesanstalt für Arbeit (BA) (2004) *Eingliederungsbilanz 2003*. Nürnberg: Bundesergebnisse.

Bundesanstalt für Arbeit (BA) (2005) *Eingliederungsbilanz 2004*. Nürnberg: Bundesergebnisse.

Bundesarbeitsgemeinschaft der Freien Wohlfahrtspflege (1995) *Alte Migranten in Deutschland: Wachsende Herausforderungen an Migrationssozialarbeit und Altenhilfe*. Bonn: KDA.

Bundesarbeitsgemeinschaft der Senioren-Organisationen (ed.) (2004) *Stellungnahme der Expertengruppe Nationaler Aktionsplan: Zur Erarbeitung eines Aktionsplans durch die Bundesregierung zur Bewältigung der demografischen Herausforderung*. www.bagso.de/fileadmin/NAP/stellungnahme_nap.pdf, accessed 4 July 2008.

Bundesministerium für Arbeit und Soziale Sicherung (ed.) (2006) *Übersicht über das Sozialrecht, Stand Januar 2006*. Bonn. www.bmas.bund.de/BMAS/Redaktion/Pdf/Publikationen/Uebersicht-ueber-das-Sozialrecht/2006, accessed 2 February 2007.

Bundesministerium für Arbeit und Sozialordnung (2001) *Übersicht über das Sozialrecht*. Bonn.

Bundesministerium für Familie, Senioren, Frauen and Jugend (BMFSFJ) (ed.) (2001) *Alter und Gesellschaft: Dritter Altenbericht: Stellungnahme der Bundesregierung*. Berlin.

Bundesministerium für Familie, Senioren, Frauen and Jugend (BMFSFJ) (ed.) (2003) *Die Familie im Spiegel der amtlichen Statistik*. Berlin.

Bundesministerium für Familie, Senioren, Frauen und Jugend (BMFSFJ) (2005a) *Mit der steigenden Wirtschaftskraft Älterer rechnen*. Pressemitteilung, 13 August. Bonn and Berlin.

Bundesministerium für Familie, Senioren, Frauen und Jugend (BMFSFJ) (ed.) (2005b) *Möglichkeiten und Grenzen selbständiger Lebensführung in Privathaushalten: Ergebnisse der Studie, MuG III*. Berlin.

Bundesministerium für Familie, Senioren, Frauen und Jugend (BMFSFJ) (2006a) *Fünfter Bericht zur Lage der älteren Generation in der Bundesrepublik Deutschland*. Berlin.

Bundesministerium für Familie, Senioren, Frauen und Jugend (BMFSFJ) (2006b) *Potenziale des Alters in Wirtschaft und Gesellschaft: Der Beitrag älterer Menschen zum Zusammenhalt der Generationen. Fünfter Bericht zur Lage der älteren Generation in der Bundesrepublik Deutschland* ('Fifth Altenbericht'). Berlin.

Bundesministerium für Familie, Senioren, Frauen und Jugend (BMFSFJ) and Matthäi, I. (2004) *Lebenssituation der älteren alleinstehenden Migrantinnen*. Berlin.

Bundesministerium für Gesundheit und Soziale Sicherung (BMGS) (2005) *Lebenslagen in Deutschland: Der 2. Armuts- und Reichtumsbericht der Bundesregierung*. Bonn.

Bundesregierung (1993) *Antwort der Bundesregierung auf die große Anfrage der Abgeordneten Andres, Gilges, Hämmerle, weiterer Abgeordneter und der Fraktion der SPD – Drucksache 12/4009 – Situation ausländischer Rentner und Senioren in der Bundesrepublik Deutschland*. Bundestagsdrucksache 12/5796. Bonn.

Bundesregierung (1997) *Bericht der Beauftragten der Bundesregierung für Ausländerfragen zur Lage der Ausländer in der Bundesrepublik Deutschland*. Bundestagsdrucksache 13/9484. Bonn.

Bundesregierung (2003) *Perspektiven für Deutschland. Unsere Strategie für eine nachhaltige Entwicklung*. Berlin.

Bundesregierung (2004) *Erster Armuts- und Reichtumbericht: Bundestagsdrucksache*. Berlin.

Bundesregierung (2005) *Bericht der Beauftragten der Bundesregierung für Migration, Flüchtlinge und Integration*. Drucksache 15/5829. Berlin.

Burmeister, J. (2006) 'Ältere Menschen als "seniorTrainerinnen" – Das Modellprogramm "Erfahrungswissen für Initiativen" (EFI)', in Schröter, K. and Zängl, P. (eds) *Altern und bürgerschaftliches Engagement: Aspekte der Vergemeinschaftung und Vergesellschaftung in der Lebensphase Alter*. Wiesbaden: VS Verlag für Sozialwissenschaften, pp. 245–59.

Busse, R., Schlette, S. and Weinbrenner, S. (2005), *Health Policy Developments Issue 4: Focus on Access, Primary Care, Health Care Organization*. Gütersloh. www.bertelsmann-stiftung.de/publications, accessed 10 December 2006.

Busse, R., Zentner, A. and Schlette, S. (2006) *Gesundheitspolitik in Industrieländern Ausgabe 5: Im Blickpunkt: Privatisierungstrends, Patientensicherheit, Lebensstil*. Gütersloh. www.bertelsmann-stiftung.de/bst/de/media/xcms_bst_dms_16344_16345_2.pdf, accessed 10 December 2006.

Butt, J. and Mirza, K. (1996) *Social Care and Black Communities*. London: HMSO.

Butt, J. and O'Neill, A. (2004) *'Let's Move On' – Black and Minority Ethnic Older People's Views on Research Findings*. York: Joseph Rowntree Foundation.

Bytheway, B. (1997) 'Talking About Age: The Theoretical Basis of Social Gerontology', in Jamieson, A., Harper, S. and Victor, C. (eds) *Critical Approaches to Ageing and Later Life*, Buckingham: Open University Press.

Cabinet Office (2000) *Winning the Generation Game – Improving Opportunities for People Aged 50–65 in Work and Community Activity*. London.

Cabinet Office (2001) *Towards Equality and Diversity: Implementing the Employment and Race Directives*. London.

Calasanti, T. (2003) 'Masculinities and Care Work in Old Age', in Arber, S., Davidson, K. and Ginn J. (eds) *Gender and Ageing: Changing Roles and Relationships*, Maidenhead: Open University Press, pp. 15–30.

Callahan, D. (1987) *Setting Limits: Medical Goals in an Ageing Society*, New York: Simon and Schuster.

Callahan, D. (1996), 'Ageing and the Allocation of Resources', in Oberender, P. (ed.) *Alter und Gesundheit*, Baden-Baden: Nomos, pp. 83–92.

CancerBACUP (2004) *Beyond the Barriers: Providing Cancer Information and Support for Black and Minority Ethnic Communities*. London: CancerBACUP.

Cancian, F. (1987) *Love in America: Gender and Self Development*. Cambridge and New York: Cambridge University Press.

Carter, S. and Beresford, P. (2000) *Age and Change: Models of Involvement for Older People*. York: Joseph Rowntree Foundation.

Chau, R. and Yu, S. (2000) 'Chinese Older People in Britain: Double Attachment to Double Detachment', in Warnes, A., Warren, L. and Nolan, M. (eds) *Care Services for Later Life: Transformations and Critiques*, London: Jessica Kingsley, pp. 259–72.

Christlich Demokratische Union / Christlich Soziale Union (CDU/CSU) (eds) (2004) *Reform der gesetzlichen Krankenversicherung – Solidarisches Gesundheitsprämien-Modell*.

Cirkel, M., Hilbert, J. and Schalk, C. (2004) *Produkte und Dienstleistungen für mehr Lebensqualität im Alter*. Expertise für den 5. Altenbericht der Bundesregierung. Gelsenkirchen: Institut für Arbeit und Technik.

Clark, H., Dyer, S. and Horwood, J (1998) *'That Bit of Help': The High Value of Low Level Preventative Services for Older People*. Bristol: Policy Press.

Clemens, W. (1992) *Arbeit – Leben – Rente: Biographische Erfahrungen von Frauen bei der Deutschen Bundespost*. Bielefeld: Kleine.

Clemens, W. (1993) 'Soziologische Aspekte eines "Strukturwandels des Alters"', in Naegele, G. and Tews, H. P. (eds) *Lebenslagen im Strukturwandel des Alters*, Opladen: Westdeutscher, pp. 61–81.

Clemens, W. and Naegele, G. (2004) 'Lebenslagen im Alter', in Kruse, A. and Martin, M. (eds) *Enzyklopädie der Gerontologie*, Bern: Hans Huber, pp. 387–402.

Coleman, D. (1996) 'Großbritannien und die internationale Migration: Die Bilanz hat sich geändert', in Fassmann, H. and Münz, R. (eds) *Migration in Europa: Historische Entwicklung, aktuelle Trends, politische Reaktionen*. Frankfurt am Main: Campus, pp. 53–89.

Comas-Herrera, A., Wittenberg, R. and Pickard, L. (2004) 'Long-term Care for Older People in the United Kingdom: Structure and Challenges', in Knapp, M., Challis, D., Fernández, J.-L. and Netten, A. (eds) *Long-Term Care: Matching Resources and Needs*, Aldershot: Ashgate, pp. 17–34.

Comptroller and Auditor General (2003) *Developing Effective Services for Older People*. HC 518 Session 2002–2003. London: The Stationery Office.

Commission for Racial Equality (1997) *Race, Culture and Community Care: An Agenda for Action*. London: CRE.

Commission for Racial Equality (2000) *The Race Relations (Amendment) Bill: Questions and Answers*. London: CRE.

Commission for Racial Equality (2002) *The Duty to Promote Race Equality: A Guide for Public Authorities (Non-statutory)*. London: CRE.

Connidis, I. A. (2001) *Family and Aging*. Thousand Oaks, CA: Sage.

Cook, J., Maltby, T. and Warren, L. (2004) 'A Participatory Approach to Understanding Older Women's Quality of Life', in Walker, A. and Hagan Hennessy, C. (eds) *Growing Older: Quality of Life in Old Age*, Maidenhead: Open University Press, pp. 149–66.

Cooper, H., Arber, S., Fee, L. and Ginn, J. (1999) *The Influence of Social Support and Social Capital on Health: A Review and Analysis of British Data*. London: Health Education Authority.

Council of the European Union (2004) *Joint Report by the Commission and the Council on Social Inclusion*. Brussels. Retrieved from europa.eu. int/comm/employment_social/soc-prot/soc-incl/final_joint_inclusion_report_2003_en.pdf.

Cumming E., and Henry, W. (1961) *Growing Old: The Process of Disengagement.* New York: Basic Books.

Daatland, S. O. and Herlofson, K. (2003) ' "Lost Solidarity" or "Changed Solidarity": A Comparative European View of Normative Family Solidarity', *Ageing and Society*, 23(5), pp. 537–60.

Davidson. K. (2001) 'Late Life Widowhood, Selfishness and New Partnership Choices: A Gender Perspective', *Ageing and Society*, 21(3), pp. 279–317.

Davidson, K., Daley, T. and Arber, S. (2003) 'Exploring the World of Older Men', in Arber, S., Davidson, K. and Ginn J. (eds) *Gender and Ageing: Changing Roles and Relationships*, Maidenhead: Open University Press, pp. 168–85.

Davis Smith, J. (1997) *The 1997 National Survey on Volunteering*, London: Institute for Volunteering.

de Jong Gierveld, J. (1999) 'A Review of Loneliness: Concepts and Definitions, Causes and Consequences', *Reviews in Clinical Gerontology*, 8, pp. 73–80.

Deeming, C. and Keen, J. (2003) 'A Fair Deal for Care in Old Age? Public Attitudes Towards the Funding of Long-term Care', *Policy and Politics*, 31(4), pp. 431–46.

Delsen, L. and Reday-Mulvey, G. (1996) *Gradual Retirement in the OECD Countries: Macro and Micro Issues and Policies*. Aldershot: Dartmouth.

Department for Education and Employment (DfEE) (2000) *Design of New Deal 50 Plus*. London.

Department for Education and Employment (DfEE) (2001) *Training Older People*. Sheffield.

Department of Employment, Training and Rehabilitation (DETR) (1998) *Modern Local Government: In Touch with the People*. London: The Stationery Office.

Department of Health (2000) *The NHS Plan*. London. www.nhsia.nhs.uk/nhsplan/ nhsplan.htm, accessed 20 November 2006.

Department of Health (2001a) *National Service Framework for Older People: Executive Summary*. London.

Department of Health (2001b), *National Service Framework for Older People*. London. www.dh.gov.uk/en/Publicationsandstatistics/Publications/Publications PolicyAndGuidance/DH_4003066, accessed 20 November 2006.

Department of Health (2001c) *National Service Frameworks*. www.dh.gov.uk/ PolicyAndGuidance/HealthAndSocialCareTopics/HealthAndSocialCareArticle/ fs/en?CONTENT_ID=4070951andchk=W3ar/W, accessed January 2005.

Department of Health (2002) *Developing Services for Minority Ethnic Older People: The Audit Tool*. Practice Guidance for Councils with Social Services Responsibilities, in support of 'From Lip Service to Real Service' (Department of Health, 2001). London.

Department of Health (2003) *Equality Framework: Priorities for Action*. London.

Department of Health (2004) *NHS Foundation Trusts*. London. www.dh.gov.uk/en/ Policyandguidance/Organisationpolicy/Secondarycare/NHSfoundationtrusts/ index.htm, accessed 22 November 2006.

Department of Health (2005a) *Delivering Race Equality in Mental Health Care*. London.

Department of Health (2005b), *Primary Care Trust Recurrent Revenue Allocations 2006–07 and 2007–08*. London. www.dh.gov.uk/en/Publicationsandstatistics/ Lettersandcirculars/Healthservicecirculars/DH_4102980, accessed 20 February 2007.

Department of Health (2005c) *Independence, Well-being and Choice: Our Vision for the Future of Adult Social Care*. London.

Department of Health (2006), *Facts about NHS Finances*. London. www.dh.gov.uk/en/Policyandguidance/Organisationpolicy/Financeandplanning/DH_4134484, accessed 20 February 2007.

Department of Health and Social Services (ed.) (1980) *Inequalities in Health*. Report of a Research Working Group (*The Black Report*). London. www.sochealth.co.uk/history/black.htm, accessed 23 November 2006.

Department of Social Security (DSS) (1998) *A New Contract for Welfare: Partnership in Pensions*. Cmnd 4179. London: The Stationery Office.

Department for Work And Pensions (DWP) (2001) *Building on Partnership: The Government Response to the Recommendations of the Better Government for Older People Programme*. London.

Department for Work and Pensions (DWP) (2002) *Simplicity, Security and Choice: Working and Saving for Retirement*. London.

Department for Work and Pensions (DWP) (2004) *Building on New Deal: Local Solutions Meeting Individual Needs*. London.

Department for Work and Pensions (DWP) (2005) *Opportunity Age – Meeting the Challenges of Ageing in the 21st Century*, London.

Der Spiegel (2005) 'Wohin mit Oma? Pflege-Notstand in Deutschland', 19, 9 May.

Deutsche Bank Research (2003a) *Demografie lässt Immobilien wackeln*. Aktuelle Themen Nr. 283. Frankfurt am Main: Rowohlt.

Deutsche Bank Research (2003b) *Pharmaceutical Market: Run on Lifestyle Drugs Boosted by Demographic Trend*. Frankfurt am Main: Rowohlt.

Deutsche Bank Research (2003c) *Demografie Spezial* (running series of contributions). Retrieved from www.dbresearch.de/servlet/reweb2.ReWEB?rwkey=u285662and%24rwframe=0).

Deutsche Bank Research (2003d) *Demografie Spezial: Auf dem Prüfstand der Senioren. Alternde Kunden fordern Unternehmen auf allen Ebenen*. Frankfurt am Main: Rowohlt.

Deutsche Bank Research (2005) *Mehr Pflegeimmobilien für eine alternde Gesellschaft*. Aktuelle Themen, Nr. 334. Frankfurt am Main: Rowohlt.

Deutscher Bundestag (ed.) (1994) *Zwischenbericht der Enquete-Kommission 'Demographischer Wandel' – Herausforderungen unserer älter werdenden Gesellschaft an den einzelnen und die Politik*. Bonn.

Deutscher Bundestag (ed.) (1998) *Demographischer Wandel: Zweiter Zwischenbericht der Enquete-Kommission 'Demographischer Wandel' – Herausforderungen unserer älter werdenden Gesellschaft an den einzelnen und die Politik*. Bonn.

Deutscher Bundestag (2002) *Vierter Bericht zur Lage der aelteren Generation in der Bundesrepublik Deutschland*. Drucksache 14/8822. Berlin.

Deutscher Bundestag (2004) *Dritter Bericht über die Entwicklung der Pflegeversicherung*. Drucksache 15/4125. Berlin.

Deutscher Verein (1998) 'Zur zukünftigen Rolle der Kommunen in der Altenhilfe (Empfehlungen)', *NDV*, S 2–5.

Dietzel-Papakyriakou, M. (1993) *Altern in der Migration: Die Arbeitsmigranten vor dem Dilemma: zurückkehren oder bleiben?* Stuttgart: Ferdinand Enke.

Dilnot, A., Kay, J. and Morris, C. (1984) *The Reform of Social Security*. Oxford: Oxford University Press.

Disney, R. and Johnson, P. (1997) 'The UK: A Working System of Minimum Pensions?', paper presented to conference on Redesigning Social Security, Kiel Institute of World Economics, Kiel, 26–27 June.

Dixon, S. (2003) 'Implications of Population Ageing for the Labour Market', in *Labour Market Trends*, London: Office for National Statistics, pp. 67–76.

Döring, D., Dudenhöffer, B. and Herdt, J. (2005) *Europäische Gesundheitssysteme unter Globalisierungsdruck: Vergleichende Betrachtung der Finanzierungsstrukturen und Reformoptionen in den EU-15 Staaten und der Schweiz.* Studie im Auftrag der Hans-Böckler-Stiftung, Report 689. Wiesbaden: Hessenagentur.

Dow, J. (2003) 'Continuing Health Care – Where Now in the Continuum of Care?', *Journal of Integrated Care*, 11(6), pp. 43–6.

Dronia, I. (2000) *Aspekte der Lebenslage älterer Spätaussiedler in Deutschland.* Diplomarbeit der Universität Dortmund, Fakultät für Erziehungswissenschaft.

DZA, Dietzel-Papakyriakou, M. and Olbermann, E. (1998) *Wohnverhältnisse älterer Migranten: Expertisenband 4 zum Zweiten Altenbericht der Bundesregirung.* Frankfurt am Main: Campus.

Ebbinghaus, B. (2001) 'The Political Economy of Early Retirement', in Ebbinghaus, B. and Manow, P. (eds) *Comparing Welfare Capitalism: Social Policy and Political Economy in Europe, Japan and the USA.* London: Routledge, pp. 76–101.

Economic Policy Committee (2003) *The Impact of Ageing Populations on Public Finances: Overview of Analysis Carried out at EU level and Proposals for a Future Work Programme.* EPC/ECFIN/435/03. Brussels: Commission of the European Communities.

Education and Employment Committee (2001) *Government's Response to the Seventh Report from the Committee Session 2000–01: Age Discrimination in Employment.* Norwich: The Stationery Office.

Eifert, B. (2006) 'Seniorenvertretungen in Nordrhein-Westfalen als Beispiel politischer Partizipation älterer Menschen', in Schröter, K. and Zängl, P. (eds) *Altern und Bürgerschaftliches Engagement: Aspekte der Vergemeinschaftung und Vergesellschaftung der Lebensphase Alter.* Wiesbaden: VS Verlag für Sozialwissenschaften, pp. 261–83.

Enquete-Kommission 'Zukunft des Bürgerschaftlichen Engagements' (2002) *Bericht Bürgerschaftliches Engagement: auf dem Weg in eine zukunftsfähige Bürgergesellschaft.* Opladen: Deutscher Bundestag.

Esping-Andersen, G. (1990) *The Three Worlds of Welfare Capitalism.* Princeton, NJ: Princeton University Press.

Esping-Andersen, G. (1999) *Social Foundations of Postindustrial Economies.* Oxford: Oxford University Press.

European Council (2001) *Presidency Conclusions.* Stockholm, 23 and 24 March. Brussels.

European Commission (2005) *Eurostat Yearbook 2005*, Chapter 2. Brussels: European Commission.

Eurostat (2005) *Old Age Dependency Ratio.* Brussels: European Commission.

Evandrou, M. and Glaser, K. C. (2004) 'Family, Work and Quality of Life: Changing Economic and Social Roles through the Lifecourse', *Ageing and Society*, 24(5), pp. 771–91.

Evans Cuellar, A. and Wiener, J. M. (2000) 'Can Social Insurance for Long-term Care Work?' The Experience of Germany', *Health Affairs*, 19(3), pp. 8–25.

Evers, A., Leichsenring, K. and Pruckner, B. (1993) *Pflegegeld in Europa*. Vienna: European Centre for Social Welfare Policy and Research.

Expertengruppe Nationaler Aktionsplan (2004) *Zur Erarbeitung eines Aktionsplans durch die Bundesregierung zur Bewältigung der demografischen Herausforderungen.* Retrieved from www.nationaler-aktionsplan.de/veroeffentlichungen.html.

Fees, B., Martin, P. and Poon, L. (1999) 'A Model of Loneliness in Older Adults', *Journal of Gerontology, Psychological Sciences*, 54b(4), pp. 231–9.

Fachinger, U. (2001) 'Materielle Ressourcen älterer Menschen – Struktur, Entwicklung und Perspektiven', in Deutsches Zentrum für Altersfragen (ed.) *Erwerbsbiographien und materielle Lebenssituation im Alter: Expertisen zum 3. Altenbericht der Bundesregierung, Band 2*, Opladen: Leske und Budrich, pp. 131–360.

Fachinger, U. (2004) *Einkommensverwendung im Alter*. Expertise für den 5. Altenbericht der Bundesregierung. Bremen.

Fachinger, U, Oelschläger, A., and Schmähl, W. (2004) *Die Alterssicherung von Selbständigen – Bestandsaufnahme und Reformoptionen*. Beiträge zur Sozial- und Verteilungspolitik, Band 2. Münster, Hamburg, London and New York: Lit.

Falkingham, J. and Rake, K. (2001) 'Modelling the Gender Impact of British Pension Reforms', in Ginn, J., Street, D., and Arber, S. (eds) *Women, Work and Pensions*, Buckingham: Open University Press, pp. 67–85.

Fawcett Society (2005) *Black and Minority Ethnic Women in the UK*. London: Fawcett Society.

Federal Ministry of Finance (2005) *Ergebnis der 125. Sitzung des Arbeitskreises 'Steuerschätzungen' vom 10. bis 12. Mai in Berlin*. Referat I A 6, Bundesministerium der Finanzen. Berlin.

Ferring, D. and Filipp, S.-H. (1999) 'Soziale Netze im Alter', *Zeitschrift für Entwicklungspsychologie und Pädagogische Psychologie*, 31, pp. 127–37.

Filipp, S.-H. (ed.) (1981) *Kritische Lebensereignisse*. Munich: Urban und Schwarzenberg.

Filtzinger, O. (1995) 'Soziales Netz mit großen Löchern: Soziale Dienste für Migranten: Anspruch und Wirklichkeit', in Senatsverwaltung für Arbeit und Frauen Berlin (ed.) *Endstation Sehnsucht? Dkumentation der Fachtagung zu neuen Perspektiven in der Sozialarbeit mit Migrantinnen am 17. und 18. November 1994*. Berlin: Eigenverlag, pp. 25–46.

Finch, J. and Mason, J. (1993) *Negotiating Family Responsibilities*. London: Routledge.

Fiori, K., Antonucci, T. and Cortina, K. (2006) 'Social Network Typologies and Mental Health Among Older Adults', *Journal of Gerontology, Psychological Sciences*, 61b(1), pp. 25–32.

Fooken, I. (1980) *Frauen im Alter: Eine Analyse intra- und interindividueller Differenzen*. Frankfurt am Main: Peter D. Lang.

Forder, J., Knapp, M., Hardy, B., Kendall, J., Matosevic, T. and Ware, P. (2004) 'Prices, Contracts and Motivations: Institutional Arrangements in Domiciliary Care', *Policy and Politics*, 32(2), pp. 207–22.

Frankenberg, R. (1966) *Communities in Britain*. London: Penguin.

Freie und Hansestadt Hamburg, Kauth-Kokshoorn, E.-M. and Schneiderheinze, K. G. (1998) *Älter werden in der Fremde: Wohn- und Lebenssituation älterer ausländischer Hamburgerinnen und Hamburger*. Hamburg: Behörde für Arbeit, Gesundheit und Soziales.

Frerichs, F. and Kauss, T. (2001) 'Zusammenfassende Bewertung', in Ministerium für Frauen, Jugend, Familie und Gesundheit NRW (ed.) *Perspektiven der politischen Beteiligung älterer Menschen: Untersuchung zur Effektivität von Seniorenvertretungen*, Düsseldorf, pp. 107–22.

Frerichs, F. and Naegele, G. (1997a) 'Discrimination of Older Workers in Germany: Obstacles and Options for the Integration into Employment', *Journal of Ageing and Social Policy*, 9(1), pp. 89–101.

Frerichs, F. and Naegele, G. (1997b) *Public Policy Options to Assist Older Workers: The German Situation*. London: EurolinkAge.

Frerichs, F. and Naegele, G. (1999) 'Offene Altenarbeit – ein vernachlässigter Bereich der Altenpolitik in Deutschland', *Theorie und Praxis der Sozialen Arbeit*, 5, pp. 169–74.

Fries, J. (1980) 'Ageing, Natural Death and the Compression of Morbidity', *New England Journal of Medicine*, 303, pp. 130–5.

Fries, J. (2003) 'Measuring and Monitoring Success in Compressing Morbidity', *Annals of Internal Medicine*, 139, pp. 455–9.

Fries, J., Green, L. and Levine, S. (1989) 'Health Promotion and the Compression of Morbidity', *The Lancet*, 8636, pp. 481–3.

Gemeinsame Erklärung (2001) *Gemeinsame Erklärung des Bündnisses für Arbeit, Ausbildung und Wettbewerbsfähigkeit zu den Ergebnissen des 7. Spitzengespräches am 4. März 2001*. Berlin: Vervielfältigung.

Gerling, V. (1999) 'Soziale Dienstleistungen für ältere Angehörige ethnischer Minderheiten: Erfahrungen und Handlungsansätze aus der kommunalen Praxis in Leeds/GB', in Hilbert, J. and Naegele, G. (eds) *Qualifizierte Dienstleistungen: Internationale Erfahrungen und Herausforderungen für den Strukturwandel im Ruhrgebiet*, Münster: Lit.

Gerling, V. (2001) *Soziale Dienste für zugewanderte Senioren/innen: Erfahrungen aus Deutschland und Großbritannien und ein Vergleich kommunaler Praxis der Partnerstädte Dortmund und Leeds*. Books on Demand.

Gerling, V. (2004a) 'Das "Unna-Projekt" im Licht der deutschen und internationalen Projektlandschaft', in Forschungsgesellschaft für Gerontologie e.V., Kreis Unna, Der Landrat, Multikulturelles Forum Lünen e.V. (eds) *Auch Migranten werden alt! Lebenslagen und Perspektiven in Europa. Dokumentation der Fachtagung vom 30. Juni bis 1. Juli 2003 in Lünen*, Unna.

Gerling, V. (2004b) 'Das Modellprojekt aus Sicht der wissenschaftlichen Begleitforschung', in MGSFF NRW (ed.) *Integration älterer Migrantinnen und Migranten: Ergebnisse eines Modellprojekts im Kreis Unna*, Düsseldorf.

Gerling, V. (2005) *Services for Elders from Ethnic Minorities: SEEM II, Final Report*. Leeds: Leeds City Council.

Gerlinger, T. (2003) 'Gesundheitsreform in der Schweiz – ein Modell für die Reform der Gesetzlichen Krankenversicherung?', in *Jahrbuch für Kritische Medizin, Band 38. Gesundheitsreformen – internationale Erfahrungen*, Berlin: Argument, pp. 10–30.

Gesundheitsziele.de (2006) www.gesundheitsziele.de.

GfK Wirtschaftstrendforschung (2002) *50plus 2002, Berichtsband I u. II*. Nürnberg.

Gibson, H. B. (2001) *Loneliness in Later Life*. Basingstoke: Palgrave Macmillan.

Giese, R. and Wiegel, D. (2000) 'Die hausliche Pflege und die Wirksamkeit von SGB XI-Gesetzliche Qualitatssicherung aus der Perspektive der Pflegehaushalte', *Zeitschrift fur Sozialreform*, 46, pp. 1023–47.

Ginn, J. (2003) *Gender, Pensions and the Lifecourse*. Bristol: Policy Press.

Ginn, J. and Arber, S. (1993) 'Pension Penalties: The Gendered Division of Occupational Welfare', *Work, Employment and Society*, 7(1), pp. 47–70.

GKV-WSG (2007) *Gesetz zur Stärkung des Wettbewerbs in der gesetzlichen Krankenversicherung (GKV – Wettbewerbsstärkungsgesetz – GKV-WSG) vom 02.02.2007*. Bundesratsdrucksache 75/07. Berlin.

Glendinning, C. (2006) 'Direct Payments and Health', in Bornat, J. and Leece, J. (eds) *Developments in Direct Payments*, Bristol: Policy Press.

Glendinning, C. and McLaughlin, E. (1994) *Paying for Care: Lessons from Europe*. Social Security Advisory Committee, Research Paper 5. London: HMSO.

Glendinning, C., Davies, B., Pickard, L. and Comas-Herrera, A. (2004) *Funding Long-Term Care for Older People: Lessons from Other Countries*. York: Joseph Rowntree Foundation.

Glendinning, C. and Means, R. (2004) 'Rearranging the Deckchairs on the Titanic of Long-term Care – Is Organisational Integration the Answer?', *Critical Social Policy*, 24(4), pp. 435–57.

Goddard, M. and Smith, P. (2001), 'Equity of Access to Health Care Services: Theory and Evidence from the UK', in: *Social Science and Medicine*, 53, pp. 1149–62.

Godfrey, M. ,Townsend, J. and Denby, T. (2004) *Building a Good Life for Older People in Local Communities: The Experience of Ageing in Time and Place*, York: Joseph Rowntree Foundation.

Goodman, A., Myck, M. and Shephard, A. (2003) *Sharing in the Nation's Prosperity? Pensioner Poverty in Britain*. London: Institute for Fiscal Studies.

Government Actuary Department (2003) *Government Actuary's Quinquennial Review of the National Insurance Fund*. Cmnd 6008. London: GAD.

Government Actuary Department (2005) *Occupational Pension Schemes 2004*. London: GAD.

Grabka, M. (2004) *Einkommen, Sparen und intrafamiliale Transfers von älteren Menschen*. DIW-Wochenbericht 6/2004. Berlin.

Green, G., Hadjimatheou, G. and Smail, R. (1984) *Unequal Fringes: Fringe Benefits in the UK*. London: Bedford Square Press / National Council of Voluntary Organisations.

Grierson, K. (2002) *New Deal 50 Plus: Quantitative Analysis of Job Retention*. Newcastle upon Tyne: IAD Information Centre.

Gronemeyer, R. (1997) *Die Entfernung vom Wolfsrudel: Über den drohenden Krieg der Jungen gegen die Alten*. Frankfurt am Nain: Roro.

Grumbach, C., Haukipuro, K., Heikelä, M., Huisman, R., Okkonen, A. and Skilton, T. (2002) *Independent Living Market in Germany, UK, Italy, Belgium and the Netherlands*. Technology Review 133/2002. Helsinki: National Technology Agency.

GSG (1992) 'Gesetz zur Sicherung und Strukturverbesserung der gesetzlichen Krankenversicherung (Gesundheitsstrukturgesetz – GSG)', *Bundesgesetzblatt*, 21 December, p. 2266.

Han, P. (2000) *Soziologie der Migration*. Stuttgart: UTB.

Hannah, L. (1986) *Inventing Retirement*. Cambridge: Cambridge University Press.

Hardy, B., Young, R. and Wistow, G. (1999) 'Dimensions of Choice in the Assessment and Care Management Process: The Views of Older People, Carers and Care Managers', *Health and Social Care in the Community*, 7(6), pp. 483–91.

Harkin, J. and Huber, J. (2004) *Eternal Youths: How the Baby Boomers are Having their Time Again*. London: Demos.

Hartz, P. *et al.* (2002) *Moderne Dienstleistungen am Arbeitsmarkt: Vorschläge zum Abbau der Arbeitslosigkeit und zur Umstrukturierung der Bundesanstalt für Arbeit*. Berlin: Kommission für Moderne Dienstleistungen am Arbeitsmarkt.

Hassel, A. (2001) 'The Governance of the Employment–Welfare Relationship in Britain and Germany', in Ebbinghaus, B. and Manow, P. (eds) *Comparing Welfare Capitalism: Social Policy and Political Economy in Europe, Japan and the USA*. London: Routledge, pp. 146–70.

Haug, S. (1997) *Soziales Kapital: Ein kritischer Überblick über den aktuellen Forschungsstand*. Arbeitsbericht II, Nr. 15. Mannheim: Mannheimer Zentrum für Europäische Sozialforschung.

Havinghust, R. (1954) 'Flexibility and the Social Roles of the Retired', *American Journal of Sociology*, 59(2), pp. 309–11.

Help the Aged (2002) *Nothing Personal: Rationing Social Care for Older People*. London: Help the Aged.

Help the Aged (2005a) *Minority Ethnic Elders: Falls Prevention: Year One Progress Report*. London: Help the Aged.

Help the Aged (2005b) *Review of the Housing Advice Needs of Black and Minority Ethnic Elders*. London: Help the Aged.

Help the Aged (2005c) *'Who Do We Trust?'*. London: Help the Aged.

Henwood, M. (2001) *'Future Imperfect?' Report of the King's Fund Care and Support Inquiry*. London: King's Fund.

Herlofson, K. and Daatland, S. O. (2001) 'The Limits of Intergenerational Responsibility: Values and Preferences towards Elder Care in a Comparative Perspective', paper presented to the 5th European Congress of Sociology, Helsinki, August.

Hilbert, J. and Cirkel, M. (2005) 'Das Altern der Gesellschaft – die graue Zukunft bekommt Silberstreifen!', in Behrens, F., Heinze, R., Rolf, G., Hilbert, J. and Stöbe-Blossey, S. (eds) *Ausblicke auf den aktivierenden Staat: von der Idee zur Strategie*. Berlin: Sigma.

Hilbert, J. and Naegele, G. (2001) 'The Economic Power of Ageing', in Pohlmann, S. (ed.) *The Ageing of Society as a Global Challenge – German Impulses: Integrated Report of German Expert Contributions* [to the UN Commission for Social Development]. Berlin: BMFSFJ and Eigenverlag.

Hoel, M. and Saether, E. (2003) 'Public Health Care with Waiting Time: The Role of Supplementary Private Health Care', *Journal of Health Economics*, 22, pp. 599–616.

Hoff, A. (2003) *Die Entwicklung sozialer Beziehungen in der zweiten Lebenshälfte: Ergebnisse des Alterssurveys 2002*. Kurzbericht an das Bundesministeriums für Familie, Senioren, Frauen und Jugend (BMFSFJ). Berlin: Deutsches Zentrum für Altersfragen.

Hoff, A. (2005) 'Intergenerationale Familienbeziehungen im Wandel', in Tesch-Römer, C., Engstler, H. and Wurm, S (eds) *Sozialer Wandel und individuelle Entwicklung in der zweiten Lebenshälfte*. Wiesbaden: VS Verlag für Sozialwissenschaften.

Holden, C. (2002) 'British Government Policy and the Concentration of Owner-ship in Long-term Care Provision', *Ageing and Society*, 22(1), pp. 79–94.

House of Lords (2003) *Aspects of the Economics of an Ageing Population*. Norwich: The Stationery Office.

Hurst, J. and Jee-Hughes, M. (2001) *Performance Measurement and Performance Management in OECD Health Systems*. Labour Market and Social Policy Series, Occasional Paper No. 47. Paris: OECD.

Hurst, J. and Siciliani, L. (2003), 'Tackling Excessive Waiting Times for Elective Surgery: A Comparison of Policies in Twelve OECD Countries', *Health Policy*, 72(2), pp. 201–15.

Hypovereinsbank (2001a) *Silver Living: Zur Zukunft des Wohnens im Alter*. Munich: Eigenverlag.

Hypovereinsbank (2001b) *3rd Generation: Wie sich von der Alterung profitieren lässt*. Munich: Eigenverlag.

Illing, K. (ed.) (2002) *Medical Wellness und Selbstzahler: Zur Erschliessung neuer Märkte für Rehabilitations, Kurkliniken und Sanatorien*. Berlin: TDC.

Inland Revenue (2002) *Simplifying the Taxation of Pensions*. Norwich: HMSO.

Institut für Freizeitwirtschaft (IFF) (2003) *Marktchancen im Gesundheitstouris-mus: Healthcare, Anti-Ageing, Wellness- und Beauty-Urlaub bis 2010*. Munich: Eigenverlag.

Institute for Fiscal Studies (2002) *English Longitudinal Study of Ageing* (ELSA). London: IFS.

Jacobzone, S. (1999) *The Interplay of Health Policy, Incentives and Regulations in the Treatment of Ageing-related Diseases*. OECD Project on Ageing Related Diseases. Paris: OECD.

Jeffrey, M.(1997) 'Intergenerational Relationships: An Autobiographical Perspec-tive', in Jamieson, A., Harper, S. and Victor, C. (eds) *Critical Approaches to Ageing and Later Life*, Buckingham: Open University Press.

Jerrome, D. (1993) 'Intimate Relationships', in Bond, J., Coleman, P. and Peace, S. (eds) *Ageing in Society: An Introduction to Social Gerontology*, London: Sage.

Johnson, P., Conrad, C. and Thomson, D. (eds) (1989) *Workers versus Pensioners: Intergenerational Conflict in an Ageing World*. Manchester: Manchester University Press.

Johnstone, S. (2005) *Private Funding Mechanisms for Long-Term Care*. York: Joseph Rowntree Foundation.

Karlsson, M., Mayhew, L., Plumb, R. and Rickayzen, B. (2004) 'The Comparative Effects on UK Public Expenditure of Implementing Long-term Care Systems as Practised in Japan, Germany and Sweden', paper presented to the Staple Inn Actuarial Society, London.

Katbamna, S. (2005) *Perspectives on Ageing and Financial Planning for Old Age in South Asian Communities*. Leicester: Nuffield Community Care Studies Unit, University of Leicester.

Key Note (2005) *Market Assessment 2005: Alternative Health Care*. Hampton: Key Note.

Klages, T. (1999) 'Rückblick und Perspektiven der Engagementförderung im Altern', in Braun, J. and Bischoff, S. (eds) (1999) *Bürgerschaftliches Engage-ment älterer Menschen: Motive und Aktivitäten. Engagementförderung in Kom-munen – Paradigmenwechsel in der offenen Altenarbeit*. Schriftenreihe des

Bundesministeriums für Familie, Senioren, Frauen und Jugend (BMFSFJ), Bd. 184, Stuttgart, Berlin and Cologne, pp. 13–24.

Klehm, W.-R. (ed) (2002) *Das Zwar-Konzept: Animation, Moderation und existentielle Begegnung in der Gruppenarbeit mit 'Jungen Alten' Rekonstruktion und Reflexion auf der Grundlage ethnographischer Bildungsforschung.* Münster: Lit.

Klein, R. (2004) 'The First Wave of NHS Foundation Trusts', *British Medical Journal*, 328, p. 1332.

Klie, T. and Spermann, A. (2004) *Personenbezogenes Pflegebudget.* Hanover: Vincentz.

Knopf, D., Schäffter, O. and Schmidt, R. (eds) (1989) *Produktivität des Alters.* Berlin: Deutsches Zentrum für Altersfragen.

Knopf, D., Schäuble, G. and Veelken, L. (1999) 'Früh beginnen: Perspektiven für ein produktives Altern', in Niederfranke, A., Naegele, G. and Frahm, E. (eds) *Funkkolleg Altern.* Opladen: Westdeutscher, pp. 97–158.

Knuth, M., Schweer, O. and Siemes, S. (2004) *Drei Menüs – und kein Rezept? Dienstleistungen am Arbeitsmarkt in Großbritannien, in den Niederlanden und in Dänemark.* Bonn: Friedrich-Ebert-Stiftung.

Kohli, M. (1999) 'Private and Public Transfers between Generations: Linking the Family and the State', *European Societies*, 1, pp. 81–104.

Kohli, M. and Künemund, H. (eds) (2000) *Die zweite Lebenshälfte: Gesellschaftliche Lage und Partizipation im Spiegel des Alters-Survey.* Opladen: Leske und Budrich.

Kohli, M. and Künemund, H. (2003) *Das Alterssurvey: Die zweite Lebenshälfte im Spiegel repräsentativer Daten: Aus Politik und Zeitgeschichte.* Beilage zur Wochenzeitung Das Parlament, B/20/18-25.

Kohli, M., Künemund, H., Motel, A. and Szydlik, M. (2000) 'Generationenbeziehungen', In Kohli, M. and Künemund, H. (eds) *Die zweite Lebenshalfte: Gesellschaftliche Lage und Partizipation im Spiegel des Alters-Survey.* Opladen: Leske und Budrich.

Kommission Impulse für die Zivilgesellschaft (2004) *Perspektiven für Freiwilligendienste und Zivildienst in Deutschland.* Bericht der Kommission Impulse für die Zivilgesellschaft. CD-ROM. Bundesministerium für Familie, Senioren, Frauen und Jugend (BMFSFJ). Berlin.

Kott, K. and Krebs, T. (2005) 'Einnahmen und Ausgaben privater Haushalte: Ergebnisse der Einkommens- und Verbrauchsstichprobe für das erste Halbjahr 2003', *Wirtschaft und Statistik*, 2, pp. 143–57.

Krause, N., Liang, J. and Keith V. (1990) 'Personality, Social Support and Psychological Destress in Later Life', *Psychology and Ageing*, 5, pp. 315–26.

Krieger, I. (1993) 'Operationalisierung des Lebenslagenansatzes für qualitative Forschung', in Hanesch, W. (ed.) *Lebenslageforschung und Sozialberichterstattung in den neuen Bundesländern.* Düsseldorf: Hans-Böckler-Stiftung, pp. 109–31.

Kruse A. (2003) 'Stärken des Alters erkennen und nutzen', paper presented to the Bundeskongress der Arbeitsgemeinschaft SPD 60 plus, Halle, 3 September (manuscript).

Kruse, A., Gaber, E., Gereon, H., Oster, P., Re, S. and Schutz-Nieswandt, F. (2002) *Gesundheit im Alter, Gesundheits-berichterstattung des Bundes, Heft 10.* Wiesbaden. www.gbe-bund.de/gbe10, accessed 10 July 2006.

Kruse, A. and Schmitt, E. (2005) 'Zur Veränderung des Altersbildes in Deutschland', *Aus Politik und Zeitgeschichte*, 49–50, pp. 9–17.

Künemund, H. (1999) 'Entpflichtung und Produktivität des Alters', *WSI Mitteilungen*, 1, 26–31.

Künemund, H. (2000) ' "Produktive" Tätigkeiten', in Kohli, M. and Künemund, H. (eds) *Die zweite Lebenshälfte: Gesellschaftliche Lage und Partizipation im Spiegel des Alters-Surveys*, Opladen: Leske und Budrich, pp. 277–317.

Künemund, H. (2006) 'Partizipation und Engagement älterer Menschen – Perspektiven von Frauen und Männern', in Deutsches Zentrum für Altersfragen (ed.) *Gesellschaftliches und familiäres Engagement älterer Menschen als Potenzial*, Berlin: DZA-Schriftenreihe, pp. 283–431.

Künemund, H. and Hollstein, B. (2000) 'Soziale Beziehungen und Unterstützungsnetzwerke', in Kohli, M. and Künemund, H. (eds) *Die zweite Lebenshälfte: Gesellschaftliche Lage und Partizipation im Spiegel des Alters-Survey*, Opladen: Leske und Budrich, pp. 212–76.

Künemund, H. and Rein, M. (1999) 'There is More to Receiving than Needing: Theoretical Arguments and Empirical Explorations of Crowding in and Crowding out', *Ageing and Society*, 19, pp. 93–121.

Lampert, H. and Althammer, J. (2005) *Lehrbuch der Sozialpolitik*, 7th ed. Berlin, Heidelberg and New York: Springer.

Law, I. (2004) *Chinese Action Research Project: Household Needs, Public Services and Community Organisations*. Leeds: Centre for Ethnicity and Racism Studies, University of Leeds.

Lee, G. R., Peek, C. W. and Coward, R. T. (1998) 'Race Differences in Filial Responsibility Expectations among Older Parents', *Journal of Marriage and the Family*, 60, pp. 404–12.

Lehr, U. (1977) *Psychologie des Alterns*, revised ed. Heidelberg: Quelle and Meyer (1st ed. 1972).

Lehr, U. (ed.) (1978) *Seniorinnen: Zur Situation der älteren Frau*. Darmstadt: Steinkopf.

Lehr, U. (1982) 'Zur Lebenssituation von älteren Frauen in unserer Zeit', in Mohr, G., Rummel, M. and Rückert, D. (eds) *Frauen: Psychologische Beiträge zur Arbeits- und Lebenssituation*, Munich, Vienna and Baltimore, MD: Urban und Schwarzenberg, pp. 103–22.

Lehr, U. and Thomae, H. (eds) (1987) *Formen seelischen Alterns*. Stuttgart: Enke.

Lewis, J. (2001) 'Das Vereinigte Königreich: Auf dem Weg zu einem neuen Wohlfahrtsmodell unter Tony Blair', in Olk, T., Evers, A. and Heinze, R. G. (eds) *Baustelle Sozialstaat: Umbauten und veränderte Grundrisse*, Wiesbaden: Chmielorz, pp. 585–604.

Lewis, J, (2005) 'New Labour's Approach to the Voluntary Sector: Independence and the Meaning of Partnership', *Social Policy and Society*, 4(2), pp. 121–31.

Lewis, J. and Meredith, B (1988) *Daughters Who Care: Daughters Caring for Mothers at Home*. London: Routledge.

Lohmann, W. (2005) *Gesundheitsreisen, Wellness, Fitness und Kur*. Berlin: FUR.

Lorenz-Meyer, D. (2004) 'The Ambivalence of Parental Care Among Young German Adults', in Pillemar, K. and Luscher, K. (eds) *Intergenerational Ambivalences: New Perspectives on Parent–Child Relations in Later Life*, Amsterdam: Elsevier.

Lowenstein, A. and Ogg, J. (eds) *OASIS: Old Age and Autonomy: The Role of Service Systems and Intergenerational Family Solidarity, Final Report*. Haifa: Centre for Research and Study of Aging, University of Haifa.

Luscher, K. and Lettke, F. (2004) 'Intergenerational Ambivalence: Methods, Measures, and Results of the Konstanz Study', in Pillemer, K. and Luscher, K. (eds) *Intergenerational Ambivalences: Perspectives on Parent–Child Relations in Later Life,* Amsterdam: Elsevier, pp. 153–79.

Maio, G. R. *et al.* (2004) 'Ambivalence and Attachment in Family Relations', in Pillemer, K. and Luscher, K. (eds) *Intergenerational Ambivalences: Perspectives on Parent–Child Relations in Later Life,* Amsterdam: Elsevier.

Mather, C. and Loncar, D. (2005) 'Projections of Global Mortality and Burden of Disease from 2002 to 2030', *PLoS Medincine,* 3(11), pp. 2011–30. http://medicine.plosjournals.org/archive/1549–1676/3/11/pdf/10.1371_journal.pmed.0030442-S.pdf, accessed 10 February 2007.

Means, R., Richards, S. and Smith, R. (2003) *Community Care: Policy and Practice,* 3rd ed. Basingstoke: Palgrave Macmillan.

Means, R. and Smith, R. (1998) *Community Care: Policy and Practice.* Basingstoke: Macmillan (now Palgrave Macmillan).

Menning, S. (2006) 'Die Zeitverwendung älterer Menschen und die Nutzung von Zeitpotenzialen für informelle Hilfeleistungen und bürgerschaftliches Engagement', in Deutsches Zentrum für Altersfragen (ed.) *Gesellschaftliches und familiäres Engagement älterer Menschen als Potenzial,* Berlin: DZA-Schriftenreihe, pp. 433–525.

Mercer, W. (2002) 'End of the Party', *The Economist,* 2 March.

Metz, D. and Underwood, M. (2005) *Older Richer Fitter: Identifying the Customer Needs of Britain's Ageing Population.* London: Age Concern England.

Meyer, M. (2004) *Eurofamcare: National Background Report for Germany.* Bremen: Institut fur angewandte Pflegeforschung, University of Bremen. www.eurofamcare.de.

Mielck, Andreas (2000) *Soziale Ungleichheit und Gesundheit: Empirische Ergebnisse, Erklärungsansätze, Interventionsmöglichkeiten.* Bern: Hans Huber.

Ministerium für Gesundheit, Soziales, Frauen und Familie des Landes Nordrhein-Westfalen (2003) *Seniorenwirtschaft in Nordrhein-Westfalen: Ein Instrument zur Verbesserung der Lebenssituation älterer Menschen.* Düsseldorf.

Ministerium für Gesundheit, Soziales, Frauen und Familie des Landes Nordrhein-Westfalen (2004) *Alter gestaltet Zukunft: Politik für Ältere in Nordrhein-Westfalen: Rahmenbedingungen, Leitlinien 2010, Datenreport.* Düsseldorf.

Minnemann, E. (1994) *Die Bedeutung sozialer Beziehungen fur Lebenszufriedenheit im Alter.* Regensburg: Roderer.

Minnemann, E. and Lehr, U. (1994) 'Der altere Mensch in Familie und Gesellschaft', in Olbrich, E., Sames, K. and Schramm, A. (eds) *Kompendium fur Gerontologie: Ein interdisziplinares Handbuch fur Forschung, Klinik und Praxis,* Landsberg: Ecomed.

Mold, F., Fitzpatrick, J. and Roberts, J. (2005) 'Minority Ethnic Elders in Care Homes: A Review of Literature', *Age and Ageing,* 34(22), pp. 107–13.

Moss, N. and Arrowsmith, J. (2003) *A Review of 'What Works' for Clients Aged Over 50.* London: Department for Work and Pensions.

Motel-Klingebiel, A., Krause, P. and Künemund, H. (2004) *Alterseinkommen der Zukunft – eine szenarische Skizze.* Berlin: Deutsches Zentrum für Altersfragen.

Motel-Klingebiel, A. and Tesch-Römer, A. (2004) *Generationengerechtigkeit in der sozialen Sicherung: Anmerkungen sowie ausgewahlte Literatur aus Sicht der angewandten Alternsforschung.* DZA-Papier, Nr. 42. Berlin: DZA.

Müller, W. C. (1996) ' "Erfahrungswissen älterer Menschen nutzen": Die Darstellung des Gesamtprogramms eines zehnjährigen Berliner Feld-Experiments', in Schweppe, C. (ed.) *Soziale Altenarbeit: Pädagogische Arbeitsansätze und die Gestaltung von Lebensentwürfen im Alter*, Munich: Weinheim, pp. 75–86.

Murthi, M., Orszag, M. and Orszag, P. (2001) 'Administrative Costs Under a Decentralised Approach to Individual Accounts: Lessons from the United Kingdom', in Holzmann, R. and Stiglitz, J. (eds) *New Ideas about Old Age Security*, Washington, DC: World Bank and Oxford University Press.

Nachhhaltigkeitsmission (2003) *Nachhaltigkeit in der Finanzierung der Sozialen Sicherungssysteme*. Berlin: Bundesministerium für Gesundheit und Soziale Sicherung.

Naegele, G. (1991) 'Anmerkungen zur These vom "Strukturwandel des Alters" aus sozialpolitikwissenschaftlicher Sicht', *Sozialer Fortschritt*, 5–6, pp. 162–72.

Naegele, G. (1993) 'Solidarität im Alter: Überlegungen zu einer Umorientierung der Alterssozialpolitik', *Sozialer Fortschritt*, pp. 191–8.

Naegele, G. (1999) 'Strukturen der politischen Mitbestimmung älterer Menschen in Deutschland – Eine Zwischenbilanz', *Theorie und Praxis der Sozialen Arbeit*, 4, pp. 131–7.

Naegele, G. (2002) 'Active Strategies for Older Workers in Germany', in ETUI (European Trade Union Institute) (ed.) *Active Strategies for Older Workers*, Brussels: ETUI-Eigenverlag, pp. 207–45.

Naegele, G. (2004) *Zwischen Arbeit und Rente*, 2nd ed. Augsburg: Maro (1st ed. 1992).

Naegele, G. and Heinze, R. G. *et al.* (2005) *Demografischer Wandel im Ruhrgebiet: Auf der Suche nach neuen Märkten*. Essen: Projekt-Ruhr.

Naegele, G., Olbermann, E. and Dietzel-Papakyriakou, M. (1997) 'Älter werden in der Migration: Eine neue Herausforderung für die kommunale Sozialpolitik', *Sozialer Fortschritt*, pp. 81–6.

Naegele, G. and Rohleder, C. (2001) 'Burgerschaftliches Engagement und Freiwilligenarbeit im Alter – individuelle Verpflichtung oder gesellschaftliche Gestaltungsaufgabe?', *Theorie und Praxis der Sozialen Arbeit*, 11, pp. 415–21.

Naegele, G. and Schmidt, W. (1998) 'Anmerkungen zur Zukunft der Generationenbeziehungen', in Veelken, L., Gosken, E. and Pfaff, M. (eds) *Jung und Alt: Beitrage und Perspektiven zu intergenerativen Beziehungen*, Hanover: Vincentz, pp. 89–122.

Naegele, G. and Walker, A. (2002a) *Ageing and Social Policy: Britain and Germany Compared*. London: Anglo-German-Foundation.

Naegele, G. and Walker, A. (2002b) *Ageing and Social Policy: Towards an Agenda for Policy Learning Between Britain and Germany*. A Report of the Anglo-German-Foundation for the Study of Industrial Society. London: AGF.

Naegele, G. and Walker, A. (2006) *A Guide to Good Practice in Age Management*. Dublin: European Foundation for Living and Working Conditions.

Naegele, G. and Walker, A. (2007) 'Protection: Incomes, Poverty and the Reform of the Pension Systems', in Bond, J., Peace, S., Dittmann-Kohli, F. and Westerhoff, G. J. (eds) *Ageing in Society: European Perspectives on Gerontology*, 3rd edition, London: Sage.

Naschold, F., Oppen, M., Peineman, H. and Rosenow, J. (1994) 'Germany: The Concerted Transition from Work to Welfare', in Naschold, F. and de Vroom, B. (eds) *Regulating Employment and Welfare*, New York: de Gruyter, pp. 117–82.

National Institute of Adult Continuing Education (NIACE) (2003) *Cultural Diversity – Responding to the Needs of Older People from Black and Minority Ethnic Communities.* Briefing Paper. Leicester: NIACE.

Nell, K. (2000) 'Der moderne Trend zur Selbsthilfe: Kick oder Aus für's Ehrenamt?', in Zeman, P. (ed.) *Selbsthilfe und Engagement im nachberuflichen Leben,* Regensburg: Transfer, pp. 107–16.

NHS Cymru Wales (2001) *Improving Health in Wales: A Plan for the NHS with its Partners in Cardiff.* www.wales.nhs.uk/Publications/NHSStrategydoc.pdf, accessed 20 February 2007.

NHS England (2000) *NHS Core Principles.* London. www.nhs.uk/England/AboutTheNhs/CorePrinciples.cmsx, accessed 2 December 2006.

NHS Scotland (2003) *Partnership for Care: Performance Incentive Framework.* www.scotland.gov.uk/Resource/Doc/46930/0013819.pdf, accessed 2 December 2006.

Nocon, A. and Pearson, M. (2000) 'The Role of Friends and Neighbours in Providing Support to Older People', *Ageing and Society,* 20(3), pp. 341–67.

OECD (2001) *Economic Outlook 2001.* Paris: OECD.

OECD (2003a) *Employment Outlook 2003: Towards More and Better Jobs.* Paris: OECD.

OECD (2003b) *Health Working Paper No. 6.* Paris: OECD.

OECD (2004a) *Ageing and Employment Policies: Finland.* Paris: OECD.

OECD (2004b) *Ageing and Employment Policies: Japan.* Paris: OECD.

OECD (2005) *Employment Outlook 2005.* Paris: OECD.

OECD (2006) *Health Data 2006 – Frequently Requested Data.* Paris: OECD. www.oecd.org/document/16/0,2340,en_2825_495642_2085200_1_1_1_1,00.html, accessed 28 December 2006.

Office for National Statistics (ONS) (2000) *Census, Age Distribution, by Ethnic Group and Sex.* London: The Stationery Office.

Office for National Statistics (ONS) (2002) *Social Trends 32.* London: The Stationery Office.

Office for National Statistics (ONS) (2003) *Census, April 2001 and Census, April 1991.* London: The Stationery Office.

Office for National Statistics (ONS) (2004) *Older People; Health and Caring.* London: The Stationery Office.

Office for National Statistics (ONS) (2006) *Population Trends,* No. 126, Winter. www.statistics.gov.uk/downloads/theme_population/PopTrends126.pdf, accessed 25 February 2007.

Office for National Statistics (2007) *Life Expectancy (LE), Healthy Life Expectancy (HLE) and Disability-free Life Expectancy (DFLE), at Birth and Age 65, by Sex and Country, 2001.* www.statistics.gov.uk/StatBase/Product.asp?vlnk=12964, accessed 20 March 2007.

Olbermann, E. (2003) *Entwicklung innovativer Konzepte zur sozialen Integration älterer Migranten/innen: Abschlussbericht.* ISAB-Berichte aus Forschung und Praxis.

Olk, T. (2003) 'Bürgerschaftliches Engagement: Eckpunkte einer Politik der Unterstützung freiwilliger und gemeinwohlorientierter Aktivitäten in Staat und Gesellschaft', *Neue Praxis,* 33(3/4), pp. 306–25.

Olshansky, S. J., Rudberg, M. A., Carnes, B. A., Cassel, C. K. and Brody, J. A. (1991) 'Trading off Longer Life for Worsening Health: The Expansion of Morbidity Hypothesis', *Journal of Aging and Health,* 3, pp. 194–216.

OPCS (2002) *Marriage, Divorce and Adoption Statistics.* London: The Stationery Office.

Osborn, A. (1990) *Small and Beautiful.* Edinburgh: Age Concern Scotland.

Osborn, A. (1991) *Taking Part in Community Care Planning: The Involvement of User Groups, Carer Groups and Voluntary Groups.* Leeds: Nuffield Institute for Health Service Studies, University of Leeds and Age Concern Scotland.

Otto, U. (1995) *Seniorengenossenschaften: Modell für eine neue Wohlfahrtspolitik?* Opladen: Leske und Budrich.

Owen, D. (1993) *Ethnic Minorities in Great Britain: Settlement Patterns.* Coventry: Centre for Research in Ethnic Relations, University of Warwick.

Owen, D. (1996) *Towards 2001: Ethnic Minorities and the Census.* Coventry: Centre for Research in Ethnic Relations, University of Warwick.

Pahl, R. (2002) 'Towards a More Significant Sociology of Friendships', *European Journal of Sociology,* 43(3), pp. 410–23.

Parker, G. (2000) 'The Royal Commission on Long Term Care for the Elderly', in Dean, H., Sykes, R. and Woods, R. (eds) *Social Policy Review 12,* Northumbria: Social Policy Association.

Patel, N. (1994) 'Healthy Margins: Black Elders' Care – Models, Policies and Prospects', in Ahmad, W. I. U. (ed.) *Race and Health in Contemporary Britain,* Buckingham: Open University Press.

Patel, N. (ed.) (2003) *Minority Elderly Care in Europe: Country Profiles.* Leeds: Policy Research Institute on Ageing and Ethnicity.

Peace, S. (1986) 'The Forgotten Female: Social Policy and Older Women', in Phillipson, C. and Walker, A. (eds) *Ageing and Social Policy: A Critical Assessment,* Aldershot: Gower, pp. 61–86.

Pensions Commission (2004) *Pensions: Challenges and Choices.* London: The Stationery Office.

Pensions Commission (2005) *A New Pension Settlement for the Twenty-First Century: The Second Report of the Pensions Commission.* Retrieved from www.pensionscommission.org.uk/publications/2005/annrep/main-report.pdf.

Pensions Policy Institute (PPI) (2003a) *Future Government Spending on Pensions.* PPI Briefing Note No. 3. London: PPI.

Pensions Policy Institute (PPI) (2003b) *The Pensions Landscape.* London: PPI.

Pensions Policy Institute (PPI) (2005) *What Will Pensions Cost in Future?* London: PPI.

Pensions Provision Group (PPG) (1998) *We All Need Pensions – The Prospects for Pension Provision.* London: The Stationery Office.

Perrig-Chiello, P. and Höpflinger, F. (2001) *Zwischen den Generationen. Frauen und Männer im mittleren Lebensalter.* Zürich: Seismo.

Pfaff, A., Langer, B., Mamberer, F., Kern, A. and Pfaff, M. (2003) *Zuzahlungen nach dem GKV-Modernisierungsgesetz (GMG) unter Berücksichtigung von Härtefallregelungen.* Working Paper No. 253. Department of Economics, University of Augsburg. www.wiwi.uni-augsburg.de/vwl/institut/paper/253.htm, accessed July 2004.

Pfaff, A., Langer, B., Wasem, J., Gress, S. and Rothgang, H. (2005), 'Modell der Kopfpauschalen in der Schweiz', in Gress, S., Pfaff, A. and Wagner, G. (eds) *Zwischen Kopfpauschale und Bürgerprämie,* Düsseldorf: Hans-Böckler-Stiftung.

Phillips, J. (ed.) (1995) *Working Carers.* Aldershot: Avebury.

Phillipson, C., Alhaq, E., Ullah, S. and Ogg, J. (2000a) 'Bangladeshi Families in Bethnal Green, London: Older People, Ethnicity and Social Exclusion', in Warnes, A., Warren, L. and Nolan, M. (eds) *Care Services for Later Life: Transformations and Critiques*, London: Jessica Kingsley, pp. 273–90.

Phillipson, C., Bernard, M., Phillips, J. and Ogg, J (2000b) *The Family and Community Life of Older People*. London: Routledge.

Pickard, L., Wittenberg, R., Comas-Herrera, A., Davies, B. and Darton, R. (2000) 'Relying on Informal Care in the New Century? Informal Care for Elderly People in England to 2031', *Ageing and Society*, 20(6), pp. 745–72.

PRIAE (2003) *Proposal for Chinese Extra Care Home in London*. London: PRIAE.

PRIAE (2004) *Summary Findings of the Minority Elderly Care (MEC) Project*. Research Briefing. London: PRIAE.

PRIAE, Housing Learning and Improvement Network (2005) *Developing Extra Care Housing for Black and Minority Ethnic Elders: An Overview of the Issues, Examples and Challenges*. London: PRIAE.

PricewaterhouseCoopers (2006) *Generation 55+ – Chancen für Handel und Konsumgüterindustrie*. Frankfurt am Main: Rowohlt.

Public Health Care Modernisation Act (2003) 'Gesetz zur Modernisierung der gesetzlichen Krankenversicherung (GKV– Modernisierungsgesetz – GMG)', *Bundesgesetzblatt*, 19 November, p. 2190.

Putnam, R. D. (2000), *Bowling Alone: The Collapse and Revival of American Community*. New York and London: Simon and Schuster.

Qualität in der ambulanten und stationären Pflege 1 (2004) Bericht des Medizinischen Dienstes der Spitzenverbände der Krankenkassen (MDS) § 118 Abs. 4 SGB XI.

Quresli, H. and Walker, A. (1989) *The Caring Relationship*. Basingstoke: Macmillan (now Palgrave Macmillan).

Randall, S. (1991) 'Local Government and Equal Opportunities in the 1990s', *Critical Social Policy*, 11(1), pp. 38–58.

Rauschenbach, T. (1999) ' "Ehrenamt" – eine Bekannte mit (zu) vielen Unbekannten: Randnotizen zu den Defiziten der Ehrenamtsforschung', in Kistler, E., Noll, H.-H. and Priller, E (eds) *Perspektiven gesellschaftlichen Zusammenhalts*, Berlin: Sigma, pp. 67–76.

Rawlingson, K. and McKay, S. (2005) *Attitudes to Inheritance in Britain*. Bristol: Policy Press.

Reichert, A. and Born, A. (2003) *Einkommen und Ausgaben älterer Menschen in Nordrhein-Westfalen*. Ergebnisse einer Repräsentativumfrage. Düsseldorf: MGSFF NRW.

Reichert, M. (2003) 'Vereinbarkeit von Erwerbstatigkeit und Pflege: Ein Uberblick zum neuesten Forschungsstand', in Reichert, M., Maly-Lukas, N. and Schonknecht, C. (eds) *Alter werdende und altere Frauen heute – zur Vielfalt ihrer Lebenssituation*, Opladen: Westdeutscher, pp. 123–48.

Reichert, M., Carell, A., Pearson, M. and Nocon, A. (2003) *Informelle auserfamiliare Unterstutzungsnetzwerke alterer Menschen mit Hilfe- und Pflegebedarf: Eine deutsch-britische Vergleichsstudie*. Munster: Lit.

Reichert, M. and Naegele, G. (1999) 'Balancing Work and Care in Germany', in Lechner, V. and Neal, M. (eds) *Work and Caring for the Elderly: An International Perspective*, Washington, DC: Taylor and Francis.

Reichert, M. and Weidekamp-Maicher, M. (2005) 'Germany: Quality of Life in Old Age II', in Walker, A. (ed.) *Growing Older in Europe*, Maidenhead: Open University Press, pp. 159–78.

Rex, J. (1991) *Ethnic Identity and Ethnic Mobilisation in Britain*. Coventry: Centre for Research in Ethnic Relations, University of Warwick.

Richards, A. (2001) *Second Time Around – A Survey of Grandparents Raising Their Grandchildren*. London: Family Rights Group.

Richards, M. and Abas, M. (1996) 'Cross-cultural Approaches to Dementia and Depression', in Bhugra, D. and Bahl, V. (eds) *Ethnicity: An Agenda for Mental Health*, London: Gaskell.

Robinson, J. (2001) 'Long-term Care in the Twenty-first Century', in Robinson, J. (ed.) *Towards a New Social Compact for Care in Old Age*, London: King's Fund.

Robinson, R. (2002) 'NHS Foundation Trusts', *British Medical Journal*, 325, pp. 506–7.

Rohleder, C. and Bröscher, P. (2000) 'Freiwilliges Engagement Älterer – integrativ oder sozial selektiv?', in Naegele, G. and Peters, G. (eds) *Arbeit – Alter – Region: Zur Debatte um die Zukunft der Arbeit, um die demographische Entwicklung und die Chancen regionalpolitischer Gestaltung*, Dortmund: Lit, pp. 93–121.

Rohleder, C. and Bröscher, P. (2001) 'Freiwilliges Engagement älterer Menschen: Potenziale und Entwicklungsmöglichkeiten', in Ministerium für Frauen, Jugend, Familie und Gesundheit NRW (ed.) *Ausmass, Strukturen und sozialräumliche Voraussetzungen*, Düsseldorf.

Rosenbladt, B. von (1999) *Freiwilliges Engagement in Deutschland – Freiwilligensurvey 1999 – Ergebnisse der Repräsentativerhebung zu Ehrenamt, Freiwilligenarbeit und bürgerschaftlichem Engagement: Band 1: Gesamtbericht*. Schriftenreihe des Bundesministeriums für Familie, Senioren, Frauen und Jugend (BMFSFJ), Bd. 194.1. Stuttgart, Berlin and Cologne.

Rosenbrock, R. and Gerlinger, T. (2006), *Gesundheitspolitik: Eine systematische Einführung*, 2nd ed. Bern: Hans Huber.

Rosenmayr, L. and Kockeis, E. (1965) *Umwelt und Familie alter Menschen*. Neuwied am Rhein and Berlin: Luchterhand.

Rössel, G., Schäfer, R. and Wahse, J. (1999) *Alterspyramide und Arbeitsmarkt: Zum Alterungsprozess der Erwerbstätigen in Deutschland*. Frankfurt am Main: Campus.

Rothgang, H. (2003) 'Long-term Care for Older People in Germany', in Comas-Herrera, A. and Wittenberg, R. (eds) *European Study of Long-Term Care Expenditure*, Report to the European Commission, DG Employment and Social Affairs, PSSRU Discussion Paper 1840, London: LSE.

Royal College of Nursing (RCN) (no date) *Who Should Fund Long-Term Care?* London: Royal College of Nursing.

Royal Commission on Long-term Care (1999) *With Respect to Old Age: Long-term Care – Rights and Responsibilities*. London: The Stationery Office.

Rurup, B. (1999) 'Halt der Generationenvertrag? Soziale Sicherung im Alter', in Niederfranke, A., Naegele, G. and Frahm, E. (eds) *Funkkolleg Altern 2: Lebenslagen und Lebenswelten, soziale Sicherung und Altenpolitik*, Opladen: Westdeutscher.

Sachverständigenrat zur Begutachtung der Entwicklung im Gesundheitswesen (2005) *Koordination und Qualität im Gesundheitswesen: Gutachten*. Drucksache 15/5670. Berlin: Deutscher Bundestag.

Schmähl, W. (1993) 'The 1992 Reform of Public Pensions in Germany: Main Elements and some Effects', *Journal of European Social Policy*, 3(1), pp. 39–51.

Schmähl, W. (1997) 'The Public–Private Mix in Pension Provision in Germany: The Role of Employer-based Pension Arrangements and the Influence of Public

Activities', in Rein, M. and Wadensjö, E. (eds) *Enterprise and the Welfare State*, Cheltenham and Lyme, NH: Edward Elgar.

Schmähl, W. (1999) 'Pension Reforms in Germany: Major Topics, Decisions and Developments', in Müller K., Ryll, A. and Wagener, H.-J. (eds) *Transformation of Social Security: Pensions in Central-Eastern Europe*, Heidelberg: Physica, pp. 91–120.

Schmähl, W. (2000) 'Pay-As-You-Go Versus Capital Funding: Towards a More Balanced View in Pension Policy', in Hughes, G. and Stewart, J. (eds) *Pensions in the European Union: Adapting to Economic and Social Change*, Boston, MA, Dordrecht and London: Kluwer, pp. 195–208.

Schmähl, W. (2002) 'Leben die "Alten" auf Kosten der "Jungen"? Anmerkungen zur Belastungsverteilung zwischen "Generationen" in einer alternden Bevolkerung aus okonomischer Perspektive', *Zeitschrift fur Gerontologie und Geriatrie*, 35, pp. 304–14.

Schmähl, W. (2004) 'Paradigm Shift in German Pension Policy: Measures Aiming at a New Public–Private Mix and their Effects', in Rein, M. and Schmähl, W. (eds) *Rethinking the Welfare State – The Political Economy of Pension Reform*, Cheltenham and Northampton, MA: Edward Elgar, pp. 153–204.

Schmähl, W. (2005a) ' "Generationengerechtigkeit" als Begründung für eine Strategie "nachhaltiger" Alterssicherung in Deutschland', in Huber, G., Krämer, H. and Kurz, H. D. (eds) *Einkommensverteilung, Technischer Fortschritt und struktureller Wandel*, Marburg: Metropolis, pp. 441–59.

Schmähl, W. (2005b) 'Einkommenslage und Einkommensverwendungspotential Älterer in Deutschland', *Wirtschaftsdienst*, 85(3), pp. 156–65.

Schmähl, W. and Rothgang, H. (1996) 'The Long-term Costs of Public Long-term Care Insurance in Germany: Some Guesstimates', in Eisen, R. and Sloan, F. A. (eds) *Long-Term Care: Economic Issues and Policy Solutions*, Boston, MA, Dordrecht and London: Kluwer.

Schmitt, E. (2004) 'Altersbild – Begriff, Befunde und politische Implikationen', in Kruse, A. and Martin, M. (eds) *Enzyklopädie der Gerontologie: Alternsprozesse in multidisziplinärer Sicht*, Bern, Göttingen, Toronto and Seattle: Hans Huber, pp. 135–47.

Schneekloth, U. and Müller, U. (2000) *Wirkungen der Pflegeversicherung*. Baden-Baden: Nomos.

Schokkaert, E. and van Parijs, P. (2003) 'Debate on Social Justice and Pension Reform', *Journal of European Social Policy*, 13(3), pp. 245–79.

Schopf, C. and Naegele, G. (2005) 'Alter und Migration – Ein Überblick', *Zeitschrift fur Gerontologie und Geriatrie*, 6, pp. 384–95.

Schulte, A. (1995) 'Staatliche und gesellschaftliche Massnahmen gegen die Diskriminierung von Ausländern in Westdeutschland', *Aus Politik und Zeitgeschichte*, B48/95, pp. 10–21.

Schunk, M. (1998) 'The Social Insurance Model of Care for Older People in Germany', in Glendinning, C. (ed.) *Rights and Realities: Comparing New Developments in Long-Term Care for Older People*, Bristol: Policy Press.

Schunk, M. and Estes, C. (2001) 'Is German Long-term Care Insurance a Model for the United States?', *International Journal of Health Services*, 31(3), pp. 617–34.

Secretary of State for Health (2000) *The NHS Plan: The Government's Response to the Royal Commission on Long Term Care*. London: Department of Health.

Seshamani, M. and Gray, A. (2002), 'The Impact of Ageing on Expenditures in the National Health Service', *Age and Ageing*, 31(4), pp. 287–94.

SGB V (Sozialgesetzbuchs Fünftes Buch) (1988) 'Gesetzliche Krankenversicherung des Artikel 1 des Gesetzes vom 20. Dezember 1988', *Bundesgesetzblatt*, p. 2477.

Shanas, E. (1979) 'Social Myth as Hypothesis: The Case of Family Relations of Older People', *The Gerontologist*, 19, pp. 3–9.

SHARE (2005) *First Results from the Survey of Health, Ageing and Retirement in Europe*, Börsch-Supan, A. *et al.* (eds). Mannheim: Mannheim Research Institute for the Economics of Ageing.

Sheldon, S. (1948) *The Social Medicine of Old Age*. Oxford: Oxford University Press.

Siciliani, L. and Hurst, J. (2003), *Explaining Waiting Times Variations for Elective Surgery across OECD Countries*. OECD Health Working Paper No. 7. Paris: OECD.

Sinfield, A. (2002) 'The Cost and Unfairness of Pension Tax Incentives', Memorandum PEN61, in House of Commons Work and Pensions Select Committee, *The Future of UK Pensions, First Report of Session 2002–2003, Volume III*, HC 92-III, pp. 300–5.

Skapinker, M. (2005) 'The Demographic Tide Will Sweep Away Age Discrimination', *Financial Times*, 25 May.

Smith, P. K. (1995) 'Grandparenthood', in Bornstein, M. H. (ed.) *Handbook of Parenting, Vol. III*, Mahwah, NJ: Lawrence Erlbaum, pp. 89–112.

Social Services Inspectorate (1998) *They Look After Their Own, Don't They? Inspection of Community Care Services for Black and Ethnic Minority Older People*. Wetherby: Department of Health.

Soule, A., Babb, P., Evandrou, M., Balchin, S. and Zealey, L. (eds) (2005) *Focus on Older People*. Basingstoke: Palgrave Macmillan for the Office of National Statistics and Department for Work and Pensions. Retrieved from www.statistics. gov.uk/downloads/theme_compendia/foop05/Olderpeople2005.pdf.

Sozialministerium Baden-Württemberg (2002) *Aktionsprogramm Ältere Generation im Mittelpunkt*. Retrieved from www.sozialministerium.badenwuerttemberg.de/sixcms/detail.php?id=50797.

SPD (Sozialdemokratische Partei Deutschlands) (ed.) (2004) *Modell einer solidarischen Bürgerversicherung, Bericht der Projektgruppe Bürgerversicherung des SPD-Parteivorstandes*. Berlin: SPD.

Statistisches Bundesamt (ed.) (2004a) *Ausländerzentralregister*. Wiesbaden.

Statistisches Bundesamt (2004b) *Einnahmen und Ausgaben privater Haushalte: Einkommens- und Verbrauchsstichprobe 2003* (Presseexemplar). Wiesbaden.

Statistisches Bundesamt (2006a) *Gesundheit: Ausgaben, Krankheitskosten, Personal 2004* (Presseexemplar). Wiesbaden. www.destatis.de/presse/deutsch/pk/2006/gesundheit_2004i.pdf, accessed 25 November 2006.

Statistisches Bundesamt (2006b), *Leben in Deutschland: Haushalte Familien und Gesundheit 2005*. Wiesbaden. www.destatis.de/presse/deutsch/pk/2006/mikrozensus_2005i.pdf, accessed 28 December 2006.

SVRKAiG (Sachverständigenrat für die Konzertierte Aktion im Gesundheitswesen) (1996) *Gesundheitswesen in Deutschland: Kostenfaktor und Zukunftsbranche, Vol. I: Demographie, Morbidität, Wirtschaftlichkeitsreserven und Beschäftigung*. Baden-Baden: Nomos.

SVRKAiG (Sachverständigenrat für die Konzertierte Aktion im Gesundheitswesen) (2002) *Bedarfsgerechtigkeit und Wirtschaftlichkeit, Vol. I: Zielbildung, Prävention, Nutzerorientierung und Partizipation, Report 2000/2001*. Baden-Baden: Nomos.

Szinovacz, M. (1982) *Women's Retirement: Policy Implications of Recent Research.* Beverly Hills, CA, London and New Delhi: Sage.

Szydlik, M. (2000) *Lebenslange Solidaritat?* Opladen: Leske und Budrich.

Taylor, P. and Urwin, P. (1999) 'Recent Trends in the Labour Force Participation of Older People in the UK', *Geneva Papers on Risk and Insurance*, 24(4), pp. 551–79.

Taylor, P. and Walker, A. (1996) Intergenerational Relations in Employment', in Walker, A. (ed) *The New Generational Contract*, London: UCL Press, pp. 159– 86.

Taylor, P. and Walker, A. (1998) 'Employers and Older Workers: Attitudes and Employment Practices,' *Ageing and Society*, 18(3), pp. 641–58.

Tesch-Römer, C. (ed.) (2004) *Abschlussbericht Sozialer Wandel und individuelle Entwicklung in der zweiten Lebenshälfte: Ergebnisse der zweiten Welle des Alterssurveys.* Berlin: DZA.

Tews, H.-P. (1990) 'Neue und alte Aspekte des Strukturwandels des Alters', *WSI Mitteilungen*, 8, pp. 478–91.

Tews, H.-P. (1993) 'Neue und alte Aspekte des Strukturwandels des Alters', in Naegele, G. and Tews, H.-P. (eds) *Lebenslagen im Strukturwandel des Alters*, Wiesbaden: Westdeutscher, pp. 15–42.

The Economist (2002) 'Over 60 and Overlooked', 8 August.

Thomas, V. (2003) *Qualitaetsmaengel und Regelungsdefizite in der Qualitaetssicherung der ambulanten Dienste.* Berlin: Bundesministerium für Familien, Senioren, Frauen und Jugend (BMFSFJ).

Thompson, P. and Mathew, D. (2004) *Fair Enough?* London: Age Concern England.

Thornton, P. (2000) Older People Speaking Out: Developing Opportunities for Influence. York: Joseph Rowntree Foundation.

Thornton, P. and Tozer, R. (1995) *Having a Say in Change: Older People and Community Care.* York: Joseph Rowntree Foundation in association with *Community Care* magazine.

TNS-infratest (2004) *Kurzzusammenfassung – 2. Freiwilligensurvey 2004 – Ehrenamt, Freiwilligenarbeit, Bürgerschaftliches Engagement.* Retrieved from www. bmfsfj.de/ RedaktionBMFSFJ/Arbeitsgruppen/Pdf-Anlagen/2.freiwilligensurvey-kurzzusammenfassung,property=pdf.pdf.

Townsend, P. (1957) *The Family Life of Old People.* London: Routledge.

Townsend, P., Whitehead, M. and Davidson, N. (1992) *Inequalities in Health.* www.scotpho.org.uk/nmsruntime/saveasdialog.asp?lID=1057andsID=1655, accessed 28 December 2006.

United Kingdom National Report for the Open Method on Co-ordination of Health Care and Long-Term Care (2005). http://ec.europa.eu/employment_social/ social_protection/docs/hc_ltc2005_en_en.pdf, accessed 28 December 2006.

van Doorslaer, E., Masseria, C. and the OECD Health Equity Research Group Members (2004) *Income-Related Inequality in the Use of Medical Care in 21 OECD Countries.* Working Paper No. 14. Paris: OECD.

Victor, C. *et al.* (2004) 'Loneliness in Later Life', in Walker, A. and Hagan Hennessy, C. (eds) *Growing Older: Quality of Life in Old Age*, Maidenhead: Open University Press, pp. 107–26.

Viebrok, H. and Himmelreicher, R. K. (2001) *Verteilungspolitische Aspekte vermehrter privater Altersvorsorge.* Working Paper 17/2001. Bremen: Zentrum für Sozialpolitik, University of Bremen.

Vierter Bericht zur Lage der aelteren Generation in der Bundesrepublik Deutschland (2002) Drucksache 14/8822. Berlin: Deutscher Bundestag.

Vincent, J. (1999) *Politics, Power and Old Age*. Buckingham: Open University Press.

Wagner, M., Schutze, Y. and Lang, F. (1996) 'Soziale Beziehungen alter Menschen', in Mayer, K. and Baltes, P. B. (eds) *Die Berliner Altersstudie*, Berlin: Akademie, pp. 301–19.

Wagner, M. and Wolf, C. (2001) 'Alter, Familie und soziales Netzwerk', *Zeitschrift für Erziehungswissenschaft*, 4, pp. 529–54.

Walker, A. (1980) 'The Social Creation of Poverty and Dependency in Old Age', *Journal of Social Policy*, 9(1), pp. 49–75.

Walker, A. (1981) 'Towards a Political Economy of Old Age', *Ageing and Society*, 1(1), pp. 73–94.

Walker, A. (1985) 'Early Retirement: Release or Refuge from the Labour Market?', *Quarterly Journal of Social Affairs*, 1(2), pp. 211–29.

Walker, A. (1990) 'The Economic "Burden" of Ageing and the Prospect of Intergenerational Conflict', *Ageing and Society*, 10, pp. 377–96.

Walker, A. (1997) *Combating Age Barriers in Employment*. Luxembourg: Office for Official Publications of the European Communities.

Walker, A. (1998) 'Speaking for Themselves: The New Politics of Old Age in Europe', *Education and Ageing*, 13(1), pp. 13–36.

Walker, A. (2002a) 'A Strategy for Active Ageing', *International Social Security Review*, 55(1), pp. 121– 39.

Walker, A. (2002b) 'Active Strategies for Older Workers in the UK', in ETUI (European Trade Union Institute) (ed.) *Active Strategies for Older Workers*, Brussels: ETUI-Eigenverlag, pp. 403–35.

Walker, A. (ed.) (2005a) *Growing Older in Europe*. Maidenhead: McGraw-Hill.

Walker, A. and Hagan Hennessey, C. (eds) (2005) *Growing Older: Quality of Life in Old Age*. Maidenhead: McGraw-Hill.

Walker, A., Maltby, T. and Walker, C. (1997) 'Altenhilfe in Europa: Länderbericht Vereinigtes Königreich', in BMFSFJ (ed.) *Altenhilfe in Europa, Länderberichte: Dänemark, Niederlande, Vereinigtes Königreich, Frankreich, Spanien*, Stuttgart: Kohlhammer, pp. 196–323.

Walker, A. and Northmore, S. (2006) *Growing Older in a Black and Ethnic Minority Group*. London: Age Concern England.

Wanless, D. (2002) *Securing Our Future Health: Taking a Long-Term View*. London: HM Treasury.

Weekers, S. and Pijl, M. (1998) *Home Care and Care Allowances in the European Union*. Utrecht: NIZW.

Wenger, G. C. and Burholt, V. (2003) 'Changes in Levels of Social Isolation and Loneliness Among Older People in Rural Wales – 20-Year Longitudinal Study', *Canadian Journal on Aging*, 23(2), pp. 115–27.

Wenger, G. C., Davies, R., Shahtahmasebi, S. and Scott, A. (1996) 'Social Isolation and Loneliness in Old Age: Review and Model Refinement', *Ageing and Society*, 16(3), pp. 333–58.

Werner, T. (2003) 'Wählerverhalten bei der Bundestagswahl 2002 nach Geschlecht und Alter: Ergebnisse der Repräsentativen Wahlstatistik', *Wirtschaft und Statistik*, 2, pp. 171–88.

Whitton, L. (1999) *Exeter Senior Voice: Third Annual Progress Report of Exeter Age Concern's User Involvement Project.* Unpublished report. Exeter: Exeter Age Concern. Cited in Thornton (2000).

Wildner, M. and Weitkunat, R. (1998) 'Aufbau einer epidemiologisch begründeten Gesundheitsberichterstattung', *Das Gesundheitswesen*, Special Issue, pp. 11–16.

Winterbotham, M., Adams, L. and Kuechel, A. (2002) *Evaluation of the Work Based Learning for Adults – Action Research Qualitative Interviews with Employment Service Staff, Providers and Employers.* WAE128. London: Department for Work and Pensions.

Winkel, R. (2003) 'Verheerende Halbjahresbilanz bei beruflicher Weiterbildung', *Soziale Sicherheit*, 7, pp. 226–9.

World Health Organisation (WHO) (1998) *Health 21: An Introduction to the Health for all Policy Framework for the WHO European Region.* Copenhagen: WHO.

World Health Organisation (WHO) (2000) *The World Health Report 2000: Health Systems Performance.* Geneva: WHO. www.who.int/whr/2000/en/whr00_en.pdf, accessed 2 December 2006.

World Health Organisation (WHO) (2002) *Active Ageing: A Policy Framework.* Geneva: WHO.

World Health Organisation (WHO) (2003) *The World Health Report 2003, Annex 4.* Geneva: WHO. www.who.int/whr/2003/en/Annex4-en.pdf, accessed 28 December 2006.

World Health Organisation (WHO) (2006) *Highlights on Health in the United Kingdom 2004.* Copenhagen: WHO. www.euro.who.int/document/e88530.pdf, accessed 28 December 2006.

Wörz, M. and Busse, R. (2005), 'Analysing the Impact of Health-care System Change in the EU Member States – Germany', *Health Economics*, 14 (Suppl. 1), pp. 133–49.

Wouters, G. (2005) *Zur Identitätsrelevanz von freiwilligem Engagement im dritten Lebensalter.* Herbolzheim: Centaurus.

Young, M. and Willmott, P. (1957) *Family and Kinship in East London.* London: Pelican.

Zundel, I. (2006) *Kommunitarismus in einer alternden Gesellschaft: Neue Lebensentwürfe Älterer in Tauschsystemen.* Herbolzheim: Centaurus.

Zur Initiative TAURIS (2002) *In aktiv*, 1(4).

Index